PANORAMAS ET

MW01600255

# RATIONAL REPRESENTATIONS, THE STEENROD ALGEBRA AND FUNCTOR HOMOLOGY

Vincent Franjou

Eric M. Friedlander

Teimuraz Pirashvili

Lionel Schwartz

Société Mathématique de France 2003
Publié avec le concours du Centre National de la Recherche Scientifique
et du Ministère de la Culture et de la Communication

*V. Franjou*

Laboratoire Jean-Leray, Université de Nantes, 2, rue de la Houssinière, BP 92208, 44322 Nantes cedex 3, France.

*E-mail :* `franjou@math.univ-nantes.fr`

*Url :* `http://www.math.sciences.univ-nantes.fr/~franjou`

*E.M. Friedlander*

Northwestern University, Evanston IL 60208, USA.

*E-mail :* `eric@math.northwestern.edu`

*Url :* `http://www.math.nwu.edu/~eric`

*T. Pirashvili*

A.M. Razmadze Mathematical Institute, Aleksidze str. 1, Tbilisi 380093, Republic of Georgia.

*E-mail :* `pira@rmi.acnet.ge`

*E-mail :* `pira@mathematik.uni-bielefeld.de`

*Url :* `http://www.rmi.acnet.gr/~pira`

*L. Schwartz*

LAGA, Institut Galilée, Université Paris Nord, 93430 Villetaneuse, France.

*E-mail :* `schwartz@math.univ-paris13.fr`

*Url :* `http://www.math.univ-paris13.fr/`

**2000 *Mathematics Subject Classification*. —** 14L15, 18G60, 19D55, 55-02, 55S10.

***Key words and phrases*. —** Frobenius twist, Koszul complex, de Rham complex, polynomial functors, Ext-groups, group scheme, Hopf algebra, algebraic group, restricted Lie algebra, general linear group, rational representation, Schur algebra, weight, support variety, Steenrod algebra, unstable modules, K-theory, bifunctor, MacLane homology, homology of a small category, spectral sequences.

V.F.: UMR 6629 Université de Nantes/CNRS.
E.M.: Partially supported by the NSF and NSA.
T.P.: Supported by the grants INTAS-99-00817 and RTN-Network "K-theory, linear algebraic groups and related structures" HPRN-CT-2002-002.
L.S.: UMR 7539 Université Paris 13/CNRS.

# RATIONAL REPRESENTATIONS, THE STEENROD ALGEBRA AND FUNCTOR HOMOLOGY

## Vincent Franjou, Eric M. Friedlander, Teimuraz Pirashvili, Lionel Schwartz

**Abstract.** — The book presents aspects of homological algebra in functor categories, with emphasis on polynomial functors between vector spaces over a finite field. With these foundations in place, the book presents applications to representation theory, algebraic topology and $K$-theory. As these applications reveal, functor categories offer powerful computational techniques and theoretical insights.

T. Pirashvili sets the stage with a discussion of foundations. E. Friedlander then presents applications to the rational representations of general linear groups. L. Schwartz emphasizes the relation of functor categories to the Steenrod algebra. Finally, V. Franjou and T. Pirashivili present A. Scorichenko's understanding of the stable $K$-theory of rings as functor homology.

*Résumé* (**Représentations rationnelles, algèbre de Steenrod et homologie des foncteurs**)

Ce livre traite d'algèbre homologique dans les catégories de foncteurs, avec une attention particulière pour les foncteurs polynomiaux entre espaces vectoriels sur un corps fini. Il en présente des applications dans trois domaines des mathématiques : la théorie des représentations, la topologie algébrique et la $K$-théorie. À chacune de ces applications, les catégories de foncteurs apportent des avancées théoriques et des outils de calcul puissants.

D'abord, T. Pirashvili expose les bases de la théorie. E. Friedlander l'applique alors aux représentations rationnelles des groupes linéaires. L. Schwartz établit les relations de l'algèbre de Steenrod avec les catégories de foncteurs. Enfin, V. Franjou et T. Pirashvili présentent un théorème de Scorichenko : la $K$-théorie stable est l'homologie des foncteurs.

# CONTENTS

**Abstracts** ................................................................ ix

**Résumés des articles** ................................................... xi

**Introduction**, *by V. Franjou* ........................................... xiii
    Representations, polynomial representations and functors: an example ...... xiv
    Functor cohomology ................................................... xvi
    Polynomials vs polynomial maps ...................................... xvi
    Modules vs comodules ................................................xvii
    Schur algebras, hyperalgebras and the Steenrod algebra ................. xix
    Content's overview ..................................................... xx
    References ............................................................ xxi

T. PIRASHVILI — *Introduction to functor homology* ........................ 1
    0. Introduction ......................................................... 1
    1. Preliminaries ....................................................... 2
        1.1. Notations ...................................................... 2
        1.2. Frobenius twist ................................................ 2
        1.3. Invariants, coinvariants, the norm homomorphism ................. 2
        1.4. Tensors ........................................................ 3
        1.5. Divided power algebra .......................................... 4
        1.6. The Koszul complex ............................................. 5
        1.7. The de Rham complex and the Cartier isomorphism ................ 7
        1.8. Adjoint functors and Ext ....................................... 8
    2. The category $\mathcal{F}(\mathbb{K})$ ............................... 10
        2.1. Definitions and examples ...................................... 10
        2.2. Polynomial functors *à la* Eilenberg and MacLane ............... 12
        2.3. Type $(\text{FP})_\infty$ ...................................... 13
        2.4. Vanishing Lemma ............................................... 16

3. Computation of $\mathrm{Ext}^*_{\mathcal{F}}(\mathrm{Id}, S^*)$ ........................................... 17
   3.1. The main theorem ........................................ 17
   3.2. Hypercohomology spectral sequences ............................ 18
   3.3. An auxiliary complex ...................................... 19
   3.4. Proof of Theorem 3.1 ..................................... 20
4. Polynomial functors à la Friedlander and Suslin ..................... 21
   4.1. Strict polynomial functors .................................. 21
   4.2. Computations in $\mathcal{P}$ ................................... 24
References ................................................... 26

E.M. FRIEDLANDER — *Lectures on the cohomology of finite group schemes* ... 27
0. Introduction .............................................. 27
1. Affine group schemes ........................................ 28
2. Cohomological techniques ..................................... 33
3. Polynomial modules and functors ............................... 39
4. Finite generation of cohomology ................................ 43
5. Qualitative description of $\mathrm{H}^{\mathrm{ev}}(G, k)$ ......................... 47
References ................................................... 52

L. SCHWARTZ — *Algèbre de Steenrod, modules instables et foncteurs polynomiaux* 55
0. Introduction .............................................. 55
   Notations ................................................ 56
1. Le groupe additif et le dual de Milnor ............................ 56
   1.1. Le groupe additif ........................................ 56
   1.2. Le cas $p > 2$ ........................................... 58
2. Comodules sur $\mathcal{A}_p^*$, comodules instables, modules instables .............. 61
   2.1. Comodules et comodules instables ............................. 61
   2.2. Exemples fondamentaux et propriétés .......................... 62
   2.3. Produit tensoriel de (co)-modules, opérations multiplicatives ........ 64
3. Les (co)modules instables libres ................................ 65
4. Les relations d'Adem ........................................ 67
   4.1. Le théorème de Bullett et MacDonald .......................... 67
   4.2. La base de Cartan-Serre ................................... 69
   4.3. Générateurs multiplicatifs ................................. 69
5. Retour sur la condition d'instabilité ............................. 70
   5.1. Algèbres instables ....................................... 70
   5.2. Les opérations $\mathrm{Sq}_0$ et $\mathrm{P}_0$ ................................ 71
6. Sur la structure de la catégorie $\mathcal{U}$ ........................... 72
   6.1. Modules instables nilpotents, réduits et localement finis .......... 72
   6.2. La catégorie $\mathcal{U}$ est localement noethérienne ................... 73
7. Représentabilité de foncteurs .................................. 74
8. Modules de Brown-Gitler et l'algèbre de Miller ...................... 75
   8.1. Modules de Brown-Gitler ................................... 75
   8.2. L'algèbre de Miller $J_*^*$ ................................... 77
9. Modules de Carlsson ........................................ 79

10. Injectivité de H*$\mathbb{Z}/p$ et le foncteur $T$ ................................... 80
    10.1. Injectivité de H*$\mathbb{Z}/p$ ........................................... 80
    10.2. Le foncteur $T$ ....................................................... 82
    10.3. Exemples de calculs avec $t$ ........................................ 83
    10.4. Le foncteur $t$ et les produits tensoriels ............................. 83
11. Injectivité de H*$E$, la formule de Campbell et Selick, et la caractérisation
des modules nilpotents ................................................ 84
    11.1. Homomorphisme de Frobenius et action tordue des opérations de
    Steenrod .................................................................. 84
    11.2. Le scindement des modules $K(i)$ ................................... 86
    11.3. Injectivité de H*$E$ et modules nilpotents, le foncteur $t_E$ ............ 86
    11.4. $t_E$ et les produits tensoriels ...................................... 87
    11.5. Le théorème de Lannes sur les algèbres instables ................. 90
12. La catégorie $\mathcal{U}/\mathcal{N}il$ et les foncteurs analytiques ..................... 91
    12.1. L'équivalence de catégorie ......................................... 91
    12.2. L'adjoint de $f$ ...................................................... 94
    12.3. Le lemme d'annulation de Kuhn ..................................... 95
    12.4. Un lemme de finitude ............................................... 97
13. Compléments ............................................................ 98
    13.1. Le foncteur $p_n : \mathcal{U}/\mathcal{N}il \to \mathcal{M}od_{\mathbb{F}_p[\mathfrak{S}_n]}$, la filtration sur $\mathcal{U}/\mathcal{N}il$ ....... 98
    13.2. Les objets simples ................................................. 99
Références ................................................................. 99


L. Schwartz — *L'algèbre de Steenrod en topologie* ......................... 101
    1. L'algèbre de Steenrod comme algèbre des opérations cohomologiques stables 101
        1.1. La cohomologie singulière ........................................ 101
        1.2. Le théorème de représentabilité de Brown et les espaces d'Eilenberg-
        MacLane ............................................................... 102
        1.3. Les opérations cohomologiques, les relations d'Adem ............. 103
    2. La condition d'instabilité et les algèbres instables ..................... 104
        2.1. Modules instables ................................................ 104
        2.2. Algèbres instables ............................................... 104
    Références ................................................................ 105


V. Franjou & T. Pirashvili — *Stable K-theory is bifunctor homology (after
A. Scorichenko)* ........................................................... 107
    0. Introduction ........................................................... 107
    1. Homology of general linear groups and stable K-theory ................. 108
    2. Preliminaries from homological algebra ................................ 110
        2.1. Universal sequences of functors ................................. 110
        2.2. A lemma on collapsing spectral sequences ....................... 110
        2.3. Categories of functors .......................................... 111
        2.4. Tor in functor categories ....................................... 112
        2.5. Homology of small categories .................................... 114

3. Finite degree functors ..................................................... 117
    3.1. Cross-effects ......................................................... 117
    3.2. Functors of finite degree ............................................. 118
    3.3. A cancellation lemma .................................................. 119
4. Proof of Scorichenko's Theorem ............................................ 119
    4.1. Stable $K$-theory is homology of the category $\mathbb{E}$ ........................ 119
    4.2. Another cancellation lemma ........................................... 120
    4.3. The homology of the category $\mathbb{E}$ ................................. 121
5. General Linear homology with twisted coefficients ......................... 124
References ..................................................................... 126

**Index of notation** .......................................................... 127

**Index** ...................................................................... 129

**Index terminologique** ....................................................... 131

# ABSTRACTS

*Introduction to functor homology*
TEIMURAZ PIRASHVILI ........................................................ 1

The aim of these notes is to provide the reader with the computational tools of functor homology. We do not assume any prior knowledge in the subject. Therefore we recall not only the basic notions on functors, but also Koszul and de Rham complexes, Cartier's homomorphism etc. We introduce two versions of polynomial functors over finite fields and explain the computation of Ext-groups between the identity functor and (twisted) symmetric powers in both categories, due to Franjou-Lannes-Schwartz and Friedlander-Suslin repectively.

*Lectures on the cohomology of finite group schemes*
ERIC M. FRIEDLANDER ........................................................ 27

We provide an introduction to the cohomology of finite group schemes, a class of objects which includes finite groups and $p$-restricted Lie algebras. Various qualitative results, known earlier for finite groups by work of Quillen and others, are extended to this general context. Various computational techniques which arise from classical homological algebra are recalled. We then proceed to discuss the essential role of strict polynomial functors in the proof of the fundamental theorem which asserts that the cohomology of a finite group scheme is finitely generated.

*Algèbre de Steenrod, modules instables et foncteurs polynomiaux*
LIONEL SCHWARTZ ........................................................ 55

This text gives an introduction to the algebraic properties of the Steenrod algebra and of the category of unstable modules. The link with the category of polynomial functors is described.

*L'algèbre de Steenrod en topologie*
LIONEL SCHWARTZ ........................................................ 101

One recalls in this note the construction of the Steenrod algebra in homotopy theory as algebra of natural stable transformations. One gives its natural properties.

*Stable K-theory is bifunctor homology (after A. Scorichenko)*
VINCENT FRANJOU & TEIMURAZ PIRASHVILI .............................. 107

For many rings $R$, the homology with coefficients of the infinite general linear group $\mathrm{GL}(R)$ is the tensor product of its homology with trivial coefficients with another term, which has been identified as the stable $K$-theory of the ring. Scorichenko's theorem states that stable $K$-theory is functor homology.

# RÉSUMÉS DES ARTICLES

*Introduction to functor homology*
TEIMURAZ PIRASHVILI ........................................................ 1

     Ces notes se proposent de donner au lecteur des outils pour le calcul de l'homologie des foncteurs. Comme on ne lui suppose aucune connaissance préalable du sujet, on rappelle les notions de base, ainsi que les complexes de Koszul et de de Rham, l'isomorphisme de Cartier, etc. Nous introduisons deux notions de foncteur polynomial sur un corps fini, et nous expliquons pour chacune le calcul des groupes d'extensions entre le foncteur identité et les puissances symétriques, dû respectivement à Franjou-Lannes-Schwartz et Friedlander-Suslin.

*Lectures on the cohomology of finite group schemes*
ERIC M. FRIEDLANDER ........................................................ 27

     Ce texte est une introduction à la cohomologie des schémas en groupes finis. Cette classe d'objets contient les groupes finis et les algèbres de Lie restreintes. Plusieurs résultats qualitatifs, établis pour les groupes finis par Quillen et d'autres, leurs sont généralisés. On rappelle les méthodes de calcul de l'algèbre homologique, puis on explique l'intervention déterminante des foncteurs polynomiaux stricts dans la démonstration qui établit que la cohomologie d'un schéma en groupes fini est de type fini.

*Algèbre de Steenrod, modules instables et foncteurs polynomiaux*
LIONEL SCHWARTZ ........................................................ 55

     Ce texte donne une introduction aux propriétés algébriques de l'algèbre de Steenrod et de la catégorie des modules instables. Les relations avec la catégorie des foncteurs polynomiaux sont établies.

*L'algèbre de Steenrod en topologie*
LIONEL SCHWARTZ ........................................................ 101

     On rappelle dans cette note la construction de l'algèbre de Steenrod en théorie de l'homotopie comme algèbre d'opérations naturelles stables. On donne ses principales propriétés.

*Stable K-theory is bifunctor homology (after A. Scorichenko)*
VINCENT FRANJOU & TEIMURAZ PIRASHVILI ............................... 107
      Pour beaucoup d'anneaux $R$, l'homologie du groupe linéaire infini avec
   coefficients s'obtient en effectuant le produit tensoriel de son homologie avec
   coefficients triviaux par un autre terme, qui n'est autre que la $K$-théorie stable
   de l'anneau. Le théorème de Scorichenko exprime la $K$-théorie stable comme
   homologie des foncteurs.

# INTRODUCTION

*by*

Vincent Franjou

This book is a sequel to a series of lectures given in Nantes, December 12–15, 2001, for Société Mathématique de France's "État de la recherche" session. The lectures presented an overview of a few recent applications of polynomial functors to homotopy and representation theory.

The organizers' interest in polynomial functors stems from a purely topological problem: Sullivan's fixed points conjecture. The proof of the conjecture in the mid 1980's [**10**, **9**] required a detailed study of the category of unstable modules over the Steenrod algebra. It revealed [**8**] that this category is closely related to the category of polynomial functors between vector spaces over a finite field. This opened the way for a productive interaction between representation theory in positive characteristic and homotopy theory. This was a timely development, because Green's 1980 book [**7**] had just shown that many ideas developed by Schur for complex numbers [**14**] extend in any characteristic. Nevertheless, functors were not yet used as such: topologists would use the Steenrod algebra, representation theorists would use Green's Schur algebras, and both would use functors only as a conceptual framework.

In the 1990's, techniques developed in the category of polynomial functors allowed elegant cohomological calculations for finite fields [**3**]. These techniques apply to a wide scope of situations where functors appear. They were used by Suslin and Friedlander to prove [**6**] the finite generation of the cohomology of finite group schemes. This in turn allowed new and impressive homological calculations to be carried out [**2**], resulting in new insights into the cohomology of the general linear groups $\mathrm{GL}_n$ in positive characteristic, and its relations to the cohomology of the finite groups $\mathrm{GL}_n(\mathbb{F}_q)$. Finally, extending functors techniques to the (twice as good) bifunctors allowed Scorichenko to elegantly describe [**17**] Waldhausen's Stable K-theory.

This book grew out of the success of the Nantes meeting. Its purpose is not to present a detailed account of fifteen years of research, especially in a topic which claims to cross the barriers of specialization. Its purpose is to give the reader a glimpse of a few aspects of polynomial functors, and encourage her to complete her knowledge through the bibliography.

Before we embark on presenting the different chapters, we would like to give the reader a concrete feeling for the kind of objects this book deals with. Through examining an easy example of representation of the general linear group, the next few pages present a modern reading of Schur's ideas, and how they lead to functors. On the way, we explain that the same representation is often meant to be quite different objects, a fact which has been a source of confusion for many of us in the past. And because most of this book deals with polynomials over finite fields, we included a section on those, again in the hope that we can spare the reader unnecessary worries.

## Representations, polynomial representations and functors: an example

Let $k$ be a field, and let us consider homogeneous polynomials in two variables $x$ and $y$. Let the group $\mathrm{GL}_2(k)$ act on the left on $k[x, y]$ by linear substitution. Precisely, we let a $2 \times 2$ matrix $\left(\begin{smallmatrix} a & c \\ b & d \end{smallmatrix}\right)$ act on $x$ to $ax + by$ and on $y$ to $cx + dy$. This defines an action on the two-dimensional $k$-vector space of homogeneous polynomials of degree 1. Let us call $V$ this vector space. Let us denote by $S^2V$ the three-dimensional $k$-vector space of homogeneous polynomials of degree two. The action of the same matrix $\left(\begin{smallmatrix} a & c \\ b & d \end{smallmatrix}\right)$, seen in the monomial basis $(x^2, y^2, xy)$ of $S^2V$, is given by the matrix:

$$(1) \qquad \begin{pmatrix} a^2 & c^2 & ac \\ b^2 & d^2 & bd \\ 2ab & 2cd & ad + bc \end{pmatrix}.$$

This formula shows us that the action extends immediately to all (even singular) matrices. This is because the coefficients above are polynomials, and no division occurs. Note as well that the coefficients can be seen as formal polynomials, that is: elements in the polynomial algebra $k[a, b, c, d]$, where the matrix coordinates $a$, $b$, $c$, $d$ are now indeterminates. The representation $S^2V$ is then called a polynomial representation of the general linear group $\mathrm{GL}_2$ over $k$. Because all the coefficients are homogeneous polynomials of degree two, one says that $S^2V$ itself is homogeneous of degree 2. This is the point of view explained in Friedlander's lectures.

Even such a simple example might discourage the reader to look in $S^2V$ for an invariant vector, or proper invariant subspace. Rightly so, for $S^2V$ is indeed a simple representation of $\mathrm{GL}_2(k)$, at least when the characteristic of $k$ is 0. But things are different when the characteristic of the field $k$ is 2: the subspace generated in $S^2V$ by $x^2$ and $y^2$ is invariant by linear substitution (so the representation $S^2V$ is not simple). This subspace looks very much like $V$ itself, except there is a square taken whenever there is a chance. We denote the subspace spanned by $x^2$ and $y^2$ by $V^{(1)}$, and call it the Frobenius twist on $V$. The squaring map $V \to S^2(V)$ is additive but not $k$-linear (except when the characteristic two field is the prime field $\mathbb{F}_2$), and the $k$-linear map $V^{(1)} \to S^2(V)$ fixes this problem. It is called the Frobenius map.

The quotient of $S^2V$ by this subspace is easily identified through the last diagonal term of the matrix (1): it is the determinant representation, and we get a short exact sequence of equivariant maps:

$$0 \longrightarrow V^{(1)} \longrightarrow S^2V \longrightarrow \det \longrightarrow 0.$$

The next question is whether $S^2V$ splits as a direct sum of $V^{(1)}$ and of the determinant.

Let us look briefly at the case when $k$ is the prime field $\mathbb{F}_2$. In this case, the invariant polynomial $x^2 + y^2 + xy$ provides a section which splits the determinant representation off $S^2V$. However this is exceptional, and the general situation is that $S^2V$ does not split:

(i) First, the splitting does not extend when one lets the singular matrices act as well.

(ii) Second, there is no such splitting when the field $k$ is larger (*e.g.* $k = \mathbb{F}_4$).

(iii) Third, there is no such splitting of $S^2V$ as a polynomial representation of $\mathrm{GL}_2$.

(iv) Fourth, there is no such splitting for homogeneous polynomials of degree 2 in more than two variables.

In order to explain what is meant in the fourth statement, we need to explain how we allow $V$ to vary. Let $V$ now denote any vector space over $k$. We view $V$, as above, as the $k$-vector space of homogeneous polynomial of degree one in $\dim(V)$ variables. The left action of $\mathrm{GL}(V)$ on $V$ so corresponds to the above linear substitution action (the indeterminates $x$, $y$ can be seen as coordinates for a chosen basis of $V$). Homogeneous polynomials of degree $d$ form the $d$-th symmetric power $S^d(V)$. Again, the simplest case is for $d = 1$, and $S^1(V) = V$. The next step is easy: describe the $d$-th symmetric power of a $k$-linear map. The correspondance thus obtained:

$$\mathrm{Hom}_k(V, W) \longrightarrow \mathrm{Hom}_k(S^d(V), S^d(W))$$

is compatible with composition of maps. This is the property that makes $S^d$ a functor. Because $S^d$ is a covariant functor of $V$, endomorphisms of $V$ act on the left on $S^d(V)$. In this way, the action on $S^d(V)$ is described in a very concise way, and in a way that makes it independent of the choice of a basis in $V$. In doing so, we gain control on the changes we may like to perform on $S^d(V)$: change of $V$ primarily, or change of fields.

Let us go back to our example and express it in terms of functors. We are looking at the *functor* $S^2$ for a field of characteristic 2. This means that instead of studying the representation $S^2V$ for a specific vector space $V$, we now look at the compatible family of the $\mathrm{End}(V)$-representations $S^2(V)$ for all finite dimensional vector spaces $V$. The quotient $S^2(V)/V^{(1)}$ defines the second exterior power $\Lambda^2(V)$, which equals the determinant representation when $V$ is two-dimensional. All in all, we obtain a short exact sequence of natural transformations of functors:

$$0 \longrightarrow V^{(1)} \longrightarrow S^2(V) \longrightarrow \Lambda^2(V) \longrightarrow 0.$$

We retrieve the above example when $V$ is two-dimensional.

This raises new questions, for example as when such a short exact sequence of natural transformations admits a natural splitting. We have already said that this is not so in this particular case (because such a splitting would preserve the action of singular matrices as well, and this fails when $\dim V \geqslant 2$). Even if only invertible matrices in the group $\mathrm{GL}(V)$ are left to act, the question of a splitting now arises for each finite-dimensional vector space $V$ over $k$. However, already when $V$ is three-dimensional, a splitting does not exist. Of course this gives another argument against a natural splitting of the short exact sequence above. It is hoped that this explains the fourth statement above as well.

Certainly, representations of the general linear group so given by evaluating a functor between vector spaces should be especially easy to handle. The good news is that they are plenty. If you manage to name a representation, it probably is one of these — for what a better name for it than the name of the functor itself, symmetric power, exterior power, tensor power *etc.* After a détour through bifunctors, even the adjoint representation $\mathrm{gl}(V)$ fits in. Schur understood this a century ago [14], and it was highlighted by Green's presentation twenty years ago [7]. They made, however, little use of functors properties.

## Functor cohomology

We have dealt so far only with a splitting problem. This leads to the classification of short exact sequences, and more generally to the classification of longer exact sequences, which is achieved by extension groups or cohomology. Often, little is known about these groups. Using the finite generation of the cohomology algebra of a finite group, Quillen [12] obtained qualitative results in the case of the finite group $\mathrm{GL}_n(k)$ for a finite field $k$. Quillen proved also that the cohomology of $\mathrm{GL}_n(k)$ goes to 0 when $n$ goes to infinity. Following our philosophy (and if we deal with singular matrices, see [4] in this book for a precise comparison statement), this limit when $n$ goes to infinity is cohomology for functors. Quillen's cancellation result suggests that cohomology simplifies when taken for functors.

Indeed, functors' properties are very handy when it comes to cohomological calculations. Base change, for example, or tensor products behave in a very formal manner for functors, and are kept well under control. These special ingredients work magic and lead to extensive detailed cohomology calculations [3, 2]. The same properties apply, in an even simpler manner, to polynomial representations as well, computing their cohomology in terms of well described classes. This newly gained control is the key in the proof of finite generation of the cohomology algebra of a finite group scheme by Friedlander and Suslin [6] — a result which opens the way to a study of rational cohomology along Quillen's lines.

## Polynomials vs polynomial maps

When working with a finite field, and using polynomials, a little care must be taken to avoid confusion between polynomials and polynomial functions. It is well known that different polynomials can take the same value at every point of the field, thus defining the same function. This is because raising to the $q$-th power is the identity for a field with $q$ elements. On the other hand, every function on a finite field is the evaluation of a polynomial.

These statements extend to several variable functions, that is to functions between $k$-vector spaces. Let $V$ be a $k$-vector space. A polynomial with coefficients in $k$ is an element in the symmetric algebra $S^*(V^\#)$, where $V^\#$ denotes the $k$-linear dual of $V$. A homogeneous polynomial of degree $d$ with coefficients in $k$ is an element in $S^d(V^\#)$. Evaluation defines a $k$-linear map:

$$S^*(V^\#) \longrightarrow k^V$$

from polynomials to functions on $V$. This map is surjective for finite fields only. The restriction $S^d(V^\#) \to k^V$ is injective if and only if $k$ contains at least $d$ elements. Indeed, if the field $k$ has $q$ elements, the kernel of the evaluation map is generated by elements of the form $x^q - x$.

Let us now define polynomial maps — in a way which is consistent with the notion of polynomial functors that is used in this book. Polynomial functors were introduced by Eilenberg and MacLane to study higher group homology [13]. Speaking in terms of derivatives or differences, a quick formulation is the following: A map is polynomial of degree less than $d$ if it satisfies the difference equations of order $d$. This is more easily formalized as follows. Consider the $k$-linear dual of $k^V$, that is the group algebra $kV$. Let $f$ be a map from $V$ to a vector space $W$. The map $f$ uniquely extends to a $k$-linear map from $kV$ to $W$. The map $f$ is polynomial of degree less than $d$ if this extension vanishes on the $d$-th power of the augmentation ideal of the group algebra $kV$. A typical degree $d$ map is: $V \to \Gamma^d(V)$, $v \mapsto v^{\otimes d}$, whose linear extension:

$$kV \longrightarrow \Gamma^d(V)$$

is the transpose of the evaluation map $S^d(V^\#) \to k^V$ (the functors $\Gamma^d$ are discussed in [11, 1.5] in this volume).

Note that the definition of a polynomial map from $V$ only uses the abelian group structure on $V$. A degree-one polynomial map is just an additive map, which can be non-linear if the field is not prime. Consider for instance the squaring map in characteristic 2.

## Modules vs comodules

As was hinted at when dealing with our introduction's example, there are many ways to describe the action of matrices on $S^2V$. We wrote it as a map from $\mathrm{GL}_2(k)$ to $\mathrm{GL}(S^2V)$, given by nine coefficients once a basis for $S^2V$ is chosen. Of course, this data is equivalent to a left module structure over the group algebra $k\,\mathrm{GL}_2(k)$, that is

a $k$-linear map:
$$k\,\mathrm{GL}_2(k) \otimes S^2V \longrightarrow S^2V$$

which is compatible with the matrix product $k\,\mathrm{GL}_2(k) \otimes k\,\mathrm{GL}_2(k) \to k\,\mathrm{GL}_2(k)$. This provides an easy way to fit representations in a category of modules. To avoid future confusion, note that the group algebra $kG$ varies covariantly with the group $G$.

It is just as good to turn the table and look at the equivalent data of a comodule structure over the dual $k\,\mathrm{GL}_2(k)^{\#}$, that is a $k$-linear map:
$$\Delta:\ S^2V \longrightarrow k\,\mathrm{GL}_2(k)^{\#} \otimes S^2V.$$

The compatibility is now with respect to the coproduct map, transpose of the product map. The notation $k\,\mathrm{GL}_2(k)^{\#}$ means $k^{\mathrm{GL}_2(k)}$, the vector space of all functions on $\mathrm{GL}_2(k)$. Certainly, coordinates $a$, $b$, $c$, $d$ are such functions. Writing the comodule map $\Delta$ in a chosen basis is just another way of listing the same nine coefficients. This goes as follows.

$$\Delta(x^2) = a^2 \otimes x^2 + b^2 \otimes y^2 + 2ab \otimes xy$$
$$\Delta(y^2) = c^2 \otimes x^2 + d^2 \otimes y^2 + 2cd \otimes xy$$
$$\Delta(xy) = ac \otimes x^2 + bd \otimes y^2 + (ad + bc) \otimes xy.$$

If one wants to consider the coefficients as polynomials, and not just as polynomial functions, this point of view is relevant. We have a map:
$$\Delta:\ S^2V \longrightarrow k[a,b,c,d] \otimes S^2V,$$

where the notation $k[a,b,c,d]$ now means polynomials in the four indeterminates $a$, $b$, $c$, $d$. It is indeed a comodule map. Let us just specify the required co-algebra structure on $k[a,b,c,d]$. This is easy, as the comultiplication map is dual to the usual matrix multiplication:
$$\Delta(a) = a \otimes a + b \otimes c, \qquad \Delta(b) = a \otimes b + b \otimes d, \ldots$$

The resulting Hopf algebra is denoted by $k[M_2]$. (Here we use $M_2$ for $2 \times 2$-matrices. The semi-group algebra $kM_2(k)$ and the coordinate algebra $k[M_2]$ are not to be confused.) The same formulae as above make $S^2V$ a comodule over $k[M_2]$. As noted before, such comodules are called polynomial representations of the general linear group (see [**5**]).

Of course, the comodule formulation is again just a matter of taste. For, turning the table once again, a homogeneous polynomial representation of $\mathrm{GL}_n$ of degree $d$ is just a module over the algebra:
$$S(n,d) := (k[M_n]^d)^{\#},$$

where $k[M_n]^d = S^d(\mathrm{End}_k(k^n)^{\#})$. The algebra $S(n,d)$ is the Schur algebra, as described in the first pages of [**7**].

## Schur algebras, hyperalgebras and the Steenrod algebra

Recall that polynomial representations of $\mathrm{GL}(V)$ are modules over the Schur algebra

$$S(V, d) = (S^d(\mathrm{End}_k(V)^{\#})^{\#} \cong \Gamma^d(\mathrm{End}_k(V)).$$

Let us denote the symmetric group by $\mathfrak{S}_d$, and write $\Gamma^d$ as invariants under the permutation action:

$$S(V, d) \cong \mathrm{End}_{k\mathfrak{S}_d}(V^{\otimes d}).$$

This formula explains how the Schur algebras act on tensor powers $V^{\otimes d}$ (and the use of idempotents in the Schur algebras to split tensor powers). Schur algebras grow with $V$. However, it is a fact that for a given degree $d$, no new information arises by increasing the dimension $n$ of $V$ beyond the degree $d$.

Summing the Schur algebras for all degrees $d$, one obtains what is often called the hyperalgebra $\Gamma(\mathrm{End}_k(V))$ (although Schur algebra might suit it better. Hyperalgebras will not appear again in this book.) It would be nice here to allow infinite dimensional vector spaces $V$, in order to keep information on arbitrary high degrees. According to the heuristics of letting $V$ get larger, this is also more suitable for dealing with functors. This can be done in a tame manner by choosing $V$ to be a graded vector space which is finite dimensional in every degree. We then define $\mathrm{End}_k(V)$ to contain only the graded endomorphisms of $V$. The resulting hyperalgebra will then act on the tensor powers $V^{\otimes d}$ for every $d$.

For a topologist, this situation is quite ordinary. Let us describe the context when the field $k$ is the field with two elements $\mathbb{F}_2$. The cohomology ring of the real projective space, $\mathrm{H}^*(\mathbb{R}\mathrm{P}^{\infty}; \mathbb{F}_2)$, is a polynomial algebra on a generator $u$ of degree one. This is a choice of a graded vector space $V$. Another choice for $V$ is the subspace $F(1)$ of the elements in $\mathrm{H}^*(\mathbb{R}\mathrm{P}^{\infty}; \mathbb{F}_2)$ whose degree is a power of 2. For each one of these graded vector spaces, the Steenrod algebra of cohomology operations acts on it, and on its tensor powers $V^{\otimes d}$. For instance, the operation denoted $\mathrm{Sq}^1$ takes $u$ to $u^2$ and $u \otimes u$ to $u^2 \otimes u + u \otimes u^2$, $\mathrm{Sq}^2$ takes $u$ to $0$, $u^2$ to $u^4$, *etc.* (see [**16**, exemple 2.3] in this volume). This action commutes with the permutation of factors, a feature shared by the hyperalgebra. However, The Steenrod algebra action is only one-directional, increasing degrees. It should therefore be comparable to the factor $\Gamma(\mathrm{End}_k^+(V))$ of the hyperalgebra, where $\mathrm{End}_k^+(V)$ is the subspace of positive degree endomorphisms — a sort of parabolic subgroup of $\mathrm{End}_k(V)$.

We shall not elaborate on this comparison any further here. An explanation at the level of modules categories [**8**], which uses functors, is to be found in this book [**15**]. It goes as follows: Lannes' functor $T$ allows to naturally associate a functor to a module over the Steenrod algebra which satisfies the unstable condition. Two unstable modules over the Steenrod algebra define the same functor if and only if they coïncide up to F-isomorphism (that is, up to nilpotent elements, this terminology goes back to Quillen [**12**]). For example, under this natural correspondence, the inclusion $F(1) \to \mathrm{H}^*(\mathbb{R}\mathrm{P}^\infty; \mathbb{F}_2)$ is sent to the evaluation map $S^1(V) \to k^{V^\#}$.

Lannes' magic functor $T$ has a nice description in these terms as well. The functor $T$ was introduced to study fixed points of finite groups actions. It is designed so that, when $p = 2$, $T(\mathrm{H}^*(X; \mathbb{F}_2))$ computes the cohomology of the mapping space $\mathrm{Map}(\mathbb{R}\mathrm{P}^\infty, X)$ under mild assumptions on $X$, for example when $X$ is a profinite space. If an unstable module $M$ corresponds to a functor $F(V)$ under the above correspondence, $T(M)$ corresponds to the functor $F(V \oplus \mathbb{F}_2)$, and the quotient $\overline{T}(M) := T(M)/M$ corresponds to the first difference $\Delta F(V) := F(V \oplus \mathbb{F}_2)/F(V)$. Thus, polynomial functors are obtained from those unstable modules that an iterate of $\overline{T}$ cancels.

## Content's overview

The book consists of four main articles that can be read independently and of a short note.

In "Introduction to functor homology" [**11**], the theory of functors is pushed until cohomological computations for finite fields are performed. This chapter has no prerequisites but the definition of Ext groups, and, by the end, the hypercohomology spectral sequences. This material can be found in Cartan and Eilenberg's book [**1**]. Basic notions are covered in a characteristic-free setting (Frobenius twist, divided powers, De Rham and Koszul complexes *etc.*), and the different notions of polynomial functors in positive characteristic are discussed on the way.

In "Lectures on the cohomology of finite group schemes" [**5**], the relevance of functor homology to the theory of algebraic groups in positive characteristic is explained. Here again, almost no prerequisite is needed beyond usual homology theory, and we trust that the first few pages and their examples can give a concise introduction to algebraic groups. The proof of finite generation is outlined. The motivation for this theorem is to realize for the rational cohomology of finite group schemes, such as $\mathrm{GL}_n$, what Quillen achieved for the cohomology of finite groups, such as $\mathrm{GL}_n(\mathbb{F}_q)$.

In "Algèbre de Steenrod, modules instables et foncteurs polynomiaux" [**15**], many of the techniques developed for unstable modules over the Steenrod algebra (including Lannes' magic functor $T$) are given a new treatment. Of course, the relation between unstable modules and polynomial functors is established. It includes quite a few bonuses, such as a universal definition of the Steenrod algebra itself, which avoids any reference to topology. A note [**16**] has been added to present the Steenrod algebra in its original topological context.

Finally, in "Stable K-theory is bifunctor homology" [**4**], Scorichenko's solution of a long standing conjecture is given. For the conjecture to be stated, functor homology needs to be extended to two-variable functors, and to general rings. The proof illustrates that functor techniques efficiently extend to this more general setting, and provide computations. The paper is meant to make up for a few accidents, among them the fact that Scorichenko has not published his result, and the untimely closing of the French consulate in Saint-Petersburg which deprived the participants in Nantes of hearing Scorichenko's scheduled talk.

# References

[1] H. CARTAN & S. EILENBERG – *Homological Algebra*, Princeton University Press, Princeton, N.J., 1956.

[2] V. FRANJOU, E.M. FRIEDLANDER, A. SCORICHENKO & A. SUSLIN – General linear and functor cohomology over finite fields, *Ann. of Math. (2)* **150** (1999), p. 663–728.

[3] V. FRANJOU, J. LANNES & L. SCHWARTZ – Autour de la cohomologie de MacLane des corps finis, *Invent. Math.* **115** (1994), no. 3, p. 513–538.

[4] V. FRANJOU & T. PIRASHVILI – Stable $K$-theory is bifunctor homology (after A. Scorichenko), in this volume.

[5] E.M. FRIEDLANDER – Lectures on the Cohomology of finite group schemes, in this volume.

[6] E.M. FRIEDLANDER & A. SUSLIN – Cohomology of finite group schemes over a field, *Invent. Math.* **127** (1997), p. 209–270.

[7] J.A. GREEN – *Polynomial representations of* $GL_n$, Lect. Notes in Math., vol. 830, Springer-Verlag, Berlin-Heidelberg-New York, 1980.

[8] H.-W. HENN, J. LANNES & L. SCHWARTZ – The categories of unstable modules and unstable algebras over the Steenrod algebra modulo nilpotents objects, *Amer. J. Math.* **115** (1993), no. 5, p. 1053–1106.

[9] J. LANNES – Sur la cohomologie modulo $p$ des $p$-groupes abéliens élémentaires, in *Proc. Durham Symposium in Homotopy Theory 1985*, London Math. Soc. Lect. Notes Series, vol. 117, Cambridge University Press, 1987, p. 97–116.

[10] H. MILLER – The Sullivan conjecture on maps from classifying spaces, *Ann. of Math. (2)* **120** (1984), p. 39–87, Erratum: *Ibid.* **121** (1985), p. 605–609.

[11] T. PIRASHVILI – Introduction to functor homology, in this volume.

[12] D. QUILLEN – The spectrum of an equivariant cohomology ring: I, II, *Ann. of Math.* **94** (1971), p. 549–572, 573–602.

[13] S. EILENBERG AND S. MACLANE – On the groups $H(\pi, n)$. II, *Ann. of Math. (2)* **60** (1954), p. 49–139.

[14] I. SCHUR – Thesis, 1903, in *Gesammelte Abhandlungen, Band I*, Springer-Verlag, 1973, p. 5–76.

[15] L. SCHWARTZ – Algèbre de Steenrod, modules instables et foncteurs polynomiaux, in this volume.

[16] _____ , L'algèbre de Steenrod en topologie, in this volume.

[17] A. SCORICHENKO – Stable K-Theory and Functor Homology over a Ring, Thesis, Evanston, 2000.

*Panoramas & Synthèses*
**16**, 2003, p. 1–26

# INTRODUCTION TO FUNCTOR HOMOLOGY

*by*

Teimuraz Pirashvili

*Abstract*. — The aim of these notes is to provide the reader with the computational tools of functor homology. We do not assume any prior knowledge in the subject. Therefore we recall not only the basic notions on functors, but also Koszul and de Rham complexes, Cartier's homomorphism etc. We introduce two versions of polynomial functors over finite fields and explain the computation of Ext-groups between the identity functor and (twisted) symmetric powers in both categories, due to Franjou-Lannes-Schwartz and Friedlander-Suslin repectively.

*Résumé* (**Introduction à l'homologie des foncteurs**). — Ces notes se proposent de donner au lecteur des outils pour le calcul de l'homologie des foncteurs. Comme on ne lui suppose aucune connaissance préalable du sujet, on rappelle les notions de base, ainsi que les complexes de Koszul et de de Rham, l'isomorphisme de Cartier, etc. Nous introduisons deux notions de foncteur polynomial sur un corps fini, et nous expliquons pour chacune le calcul des groupes d'extensions entre le foncteur identité et les puissances symétriques, dû respectivement à Franjou-Lannes-Schwartz et Friedlander-Suslin.

## 0. Introduction

The present notes follow the lectures given in Nantes by Jean-Louis Loday and the author. The aim of this series was to introduce the computational tools of functor homology. In Section 1, we recall basic facts on Koszul and de Rham complexes. In Section 2 we introduce the category $\mathcal{F}$ of functors from the category of finite dimensional vector spaces to the category of all vector spaces. Especially important is the case when the ground field is finite. In this case we prove the Schwartz theorem on type $(\mathrm{FP})_\infty$ objects. In Section 4 we introduce another category $\mathcal{P}$, the category of strongly polynomial functors. While the category $\mathcal{F}$ is related to the representation theory of unstable modules over the Steenrod algebra (see Schwartz' lectures in this

**2000** *Mathematics Subject Classification*. — 18G60.
**Key words and phrases**. — Frobenius twist, Koszul complex, de Rham complex, polynomial functors, Ext-groups.

volume), the category $\mathcal{P}$ is related to the representation theory of the algebraic groups $\mathrm{GL}_n$ (see Friedlander's lectures in this volume). In Section 3 and Section 4 we develop computational tools for Ext-groups in both categories of functors. We closely follow [6] and [9].

# 1. Preliminaries

**1.1. Notations.** — Let $\mathbb{K}$ be a field. We let $\mathcal{V}_{\mathbb{K}}$ or simply $\mathcal{V}$ be the category of vector spaces over $\mathbb{K}$. In what follows $\otimes_{\mathbb{K}}$ and $\mathrm{Hom}_{\mathbb{K}}$ are denoted by $\otimes$ and $\mathrm{Hom}$ respectively. For any set $S$ we let $\mathbb{K}[S]$ be the free $\mathbb{K}$-module spanned by $S$. If $M$ is a module and $S$ is a subset of $M$, then we let $\langle S \rangle$ be the submodule of $M$ generated by $S$. For a vector space $V$, its dual vector space is denoted by $V^{\#}$, thus $V^{\#} := \mathrm{Hom}(V, \mathbb{K})$. If $V^* = \bigoplus_{n \in \mathbb{Z}} V^n$ is a graded vector space, then we denote by $V^{*\#}$ the *graded dual* of $V^*$, that is $V^{*\#} = \bigoplus_{n \in \mathbb{Z}} V^{n\#}$.

**1.2. Frobenius twist.** — If $\mathbb{K}$ is of characteristic $p > 0$, one lets $\Phi$ denote the Frobenius endomorphism $\mathbb{K} \to \mathbb{K}$, which assigns $x^p$ to $x$. For a vector space $V$ we let $V^{(1)}$ be the vector space obtained by extending scalars *via* $\Phi$. In other words $V^{(1)}$ is the vector space generated by $v^{(1)}$, $v \in V$ subject to the following relations:

$$(v + w)^{(1)} = v^{(1)} + w^{(1)}, \quad (\lambda v)^{(1)} = \lambda^p v^{(1)}, \ \lambda \in K.$$

One defines $V^{(r)}$, $r \geqslant 1$ by induction: $V^{(r+1)} = (V^{(r)})^{(1)}$. The Frobenius twist commutes with tensor product. Thus if $A$ is an algebra, then $A^{(1)}$ has a natural algebra structure as well. The same holds for coalgebras.

If $\Phi$ is an isomorphism (thus $\mathbb{K}$ is a perfect field), then $V^{(1)}$ equivalently can be defined as a vector space which is $V$ as an abelian group, while the action of $\mathbb{K}$ on $V$ is given by

$$(\lambda, v) \longmapsto \lambda^{1/p} v.$$

Thus if $\mathbb{K} = \mathbb{F}_p$ is a finite prime field, then $V^{(1)} = V$.

**1.3. Invariants, coinvariants, the norm homomorphism.** — Let $G$ be a group and $M$ be a $G$-module. One defines

$$M^G := \{m \in M \mid gm = m\}, \quad M_G = M/\langle gm - g \rangle, \ g \in G.$$

These vector spaces are called respectively *invariants* and *coinvariants* of $M$. For any vector space $V$ considered as a $G$-module *via* trivial action, there are natural isomorphisms

$$\mathrm{Hom}_G(M, V) \cong \mathrm{Hom}(M_G, V), \quad \mathrm{Hom}_G(V, M) \cong \mathrm{Hom}(V, M^G).$$

For any $G$-module $M$, the linear dual $M^{\#}$ is also a $G$-module *via* $(gf)(m) := f(g^{-1}m)$, $g \in G$, $m \in M$ and $f \in M^{\#}$. Moreover, one has

$$(1) \qquad\qquad\qquad (M^{\#})^G \cong (M_G)^{\#}.$$

If $G$ is a finite group, then $x \mapsto \sum_{g \in G} gx$ yields a linear map

$$N_M : M_G \longrightarrow M^G,$$

which is known as the *norm homomorphism*. The norm homomorphism is an isomorphism when $|G|$ is invertible in $\mathbb{K}$.

**1.4. Tensors.** — For a vector space $V$ we let $\mathrm{T}^n(V)$ be the $n$-fold tensor product $V \otimes \cdots \otimes V$. Thus $\mathrm{T}^1(V) = V$ and $\mathrm{T}^0(V) = \mathbb{K}$. Then the graded vector space $\mathrm{T}^*(V) = \bigoplus_{n \geqslant 0} \mathrm{T}^n(V)$ has an associative graded algebra structure induced by

$$(x_1 \otimes \cdots \otimes x_n)(y_1 \otimes \cdots \otimes y_m) = x_1 \otimes \cdots \otimes x_n \otimes y_1 \otimes \cdots \otimes y_m.$$

This algebra is called the *tensor algebra spanned by* $V$. Let us note that if $\dim(V) = n$, then $\mathrm{T}^*(V)$ is isomorphic to the algebra of noncommutative polynomials in $n$ variables. Since the tensor algebra $\mathrm{T}^*(V)$ is a free associative algebra there exists a unique algebra homomorphism $\Delta : \mathrm{T}^*(V) \to \mathrm{T}^*(V) \otimes \mathrm{T}^*(V)$ such that $\Delta(v) = 1 \otimes v + v \otimes 1$. This homomorphism makes $\mathrm{T}^*(V)$ into a Hopf algebra.

The *symmetric algebra* $\mathrm{S}^*(V) = \bigoplus_{n \geqslant 0} \mathrm{S}^n(V)$ spanned by a vector space $V$ is the quotient of $\mathrm{T}^*(V)$ by the ideal generated by $x \otimes y - y \otimes x$, $x, y \in V$. The image of $x_1 \otimes \cdots \otimes x_n$ in $\mathrm{S}^n(V)$ is denoted by $x_1 \cdots x_n$. If $\dim(V) = n$, then $\mathrm{S}^*(V)$ is isomorphic to the algebra of polynomials in $n$ variables. The symmetric algebra is a functor $\mathrm{S}^*$ from the category $\boldsymbol{\mathcal{V}}$ to the category of commutative algebras, which is the left adjoint to the forgetful functor. In particular it respects coproducts. Thus the symmetric algebra functor has the exponential property, which means that there is a natural isomorphism:

$$(2) \qquad \mathrm{S}^*(V \oplus W) \cong \mathrm{S}^*(V) \otimes \mathrm{S}^*(W).$$

Since the symmetric algebra $\mathrm{S}^*(V)$ is a free commutative algebra there exists a unique algebra homomorphism $\Delta : \mathrm{S}^*(V) \to \mathrm{S}^*(V) \otimes \mathrm{S}^*(V)$ such that $\Delta(v) = 1 \otimes v + v \otimes 1$. This homomorphism makes $\mathrm{S}^*(V)$ into a Hopf algebra. Since $\Delta(v_1 \cdots v_n) = \Delta(v_1) \cdots \Delta(v_n)$ by obvious induction on $n$ we obtain

$$\Delta(v_1 \cdots v_n) = \sum_{i+j=n} \sum_{\sigma \in \mathfrak{S}_{i,j}} v_{\sigma(1)} \cdots v_{\sigma(i)} \otimes v_{\sigma(i+1)} \cdots v_{\sigma(n)}.$$

Here $\mathfrak{S}_{i,j}$ denotes the set of $(i,j)$-shuffles. Let us recall that $\sigma \in \mathfrak{S}_n$ is called an $(i, n-i)$-shuffle if $\sigma(1) < \sigma(2) < \cdots < \sigma(i)$ and $\sigma(i+1) < \sigma(i+2) < \cdots < \sigma(n)$. In particular we have

$$(3) \qquad \Delta(v^n) = \sum_{i+j=n} \binom{n}{i} v^i \otimes v^j.$$

As a consequence, we see that the algebra of polynomials $\mathbb{K}[t_1, \ldots, t_n]$ is a Hopf algebra and

$$\Delta(t_1^{k_1} \cdots t_n^{k_n}) = \sum_{i=1}^{n} \sum_{0 \leqslant j_i \leqslant k_i} \binom{k_1}{j_1} \cdots \binom{k_n}{j_n} t_1^{j_1} \cdots t_n^{j_n} \otimes t_1^{k_1-j_1} \cdots t_n^{k_n-j_n}.$$

The *exterior algebra* $\Lambda^*(V) = \bigoplus_{n \geqslant 0} \Lambda^n(V)$ *generated by* $V$ is the quotient of $\mathrm{T}^*(V)$ by the ideal generated by the elements $x \otimes x$, where $x \in V$. The vector space $\Lambda^n(V)$ is called the *n-th exterior power* of $V$. The image of $x_1 \otimes \cdots \otimes x_n$ in $\Lambda^n(V)$ is denoted by $x_1 \wedge \cdots \wedge x_n$. Thus for any $\sigma \in \mathfrak{S}_n$ one has $\sigma_*(x_1 \wedge \cdots \wedge x_n) = (-1)^{\varepsilon(\sigma)} x_1 \wedge \cdots \wedge x_n$, where $\varepsilon : \mathfrak{S}_n \to \{-1, +1\}$ is the sign homomorphism. There is a natural isomorphism

$$\Lambda^n(V^{\#}) \cong (\Lambda^n V)^{\#},$$

given by: $(f_1 \wedge \cdots \wedge f_n)(x_1 \wedge \cdots \wedge x_n) = \det(f_i(x_j))$, for $f_1, \ldots, f_n \in V^{\#}$, $x_1, \ldots, x_n \in V$.
We have

$$\Lambda^*(V \oplus W) \cong \Lambda^*(V) \otimes \Lambda^*(W).$$

There is a unique algebra homomorphism $\Delta : \Lambda^*(V) \to \Lambda^*(V) \otimes \Lambda^*(V)$ such that $\Delta(v) = 1 \otimes v + v \otimes 1$. One easily shows that this map equips $\Lambda^*(V)$ with a graded Hopf algebra structure, provided the elements of $V$ have degree one.

**1.5. Divided power algebra.** — Let $V$ be a finite dimensional vector space. One defines $\Gamma^*(V) = \bigoplus_{n \geqslant 0} \Gamma^n(V)$ as the graded dual of $S^*(V^{\#})$. Since $S^*(V^{\#})$ is a Hopf algebra the same is true also for $\Gamma^*(V)$. This algebra is known as the *divided power algebra spanned by* $V$. By definition we have

(4) $$\Gamma^n(V^{\#}) \cong (S^n V)^{\#}.$$

We refer to this isomorphism as the duality between symmetric and divided powers. We have $S^0(V) = \mathbb{K} = \Gamma^0(V)$, $S^1(V) = V = \Gamma^1(V)$. It follows from the duality between symmetric and divided powers that $\Gamma^*(V \oplus W) \cong \Gamma^*(V) \otimes \Gamma^*(W)$.
The equality (3) shows that if $\dim(V) = n$ then $\Gamma^*(V)$ as an algebra is isomorphic to $\Gamma(t_1, \ldots, t_n)$, which is defined as follows. First we consider the subgroup $\Gamma_{\mathbb{Z}}(t)$ of the ring $\mathbb{Q}[t]$ of polynomials over the rational numbers generated by

$$t^{[n]} = \frac{t^n}{n!} \in \mathbb{Q}[t], \quad n \geqslant 1.$$

It is clear that $\Gamma_{\mathbb{Z}}(t)$ is a ring. For any field $\mathbb{K}$ one defines $\Gamma_{\mathbb{K}}(t)$, or simply $\Gamma(t)$ to be the tensor product $\mathbb{K} \otimes \Gamma_{\mathbb{Z}}(t)$. Finally,

$$\Gamma(t_1, \ldots, t_n) := \Gamma(t_1) \otimes \cdots \otimes \Gamma(t_n).$$

For example, if $\mathbb{K}$ is a field of characteristic $p > 0$, then there is an isomorphism of algebras

$$\Gamma(t) \cong \mathbb{K}[e_0, e_1, \ldots, e_h, \ldots]/(e_h^p = 0),$$

where $x_h$ is the image of $t^{[p^h]}$, $h = 0, 1, \ldots$. The alternative definition (see [**13**]) of the divided power algebra is as follows: $\Gamma^*(V)$ is the graded commutative algebra generated by elements $v^{[n]}$, $v \in V$, $n \geqslant 0$, modulo the following relations:

$$v^{[0]} = 1, \quad v^{[1]} = v,$$

$$v^{[k]}v^{[h]} = \binom{k+h}{k} v^{[k+h]}, \quad (v + u)^{[k]} = \sum_{i+j=k} v^{[i]}u^{[j]},$$

$$(vu)^{[k]} = k! v^{[k]} u^{[k]} = v^k u^{[k]} = v^{[k]} u^k, \quad (v^{[k]})^{[h]} = \frac{(kh)!}{h!(k!)^h} v^{[kh]}.$$

The symmetric group $\mathfrak{S}_n$ on $n$ letters acts on $\mathrm{T}^n(V)$ by permuting the factors:

$$\sigma_*(x_1 \otimes \cdots \otimes x_n) = (x_{\sigma^{-1}(1)} \otimes \cdots \otimes x_{\sigma^{-1}(n)}).$$

It follows from the definition that $\mathrm{S}^n(V)$ is isomorphic to the coinvariants of $\mathrm{T}^n(V)$ under this action and the isomorphism (1) shows that $\Gamma^n(V)$ is isomorphic to the invariants of this action. The norm homomorphism $\mathrm{N} : \mathrm{S}^n(V) \to \Gamma^n(V)$ is an isomorphism provided $n!$ is invertible in $\mathbb{K}$, in particular if $\mathrm{Char}(\mathbb{K}) = 0$. If characteristic of $\mathbb{K}$ is $p > 0$, then the norm homomorphism is not an isomorphism in general. For example, the following sequence is exact:

$$(5) \qquad 0 \longrightarrow V^{(1)} \overset{\Phi}{\longrightarrow} \mathrm{S}^p(V) \overset{\mathrm{N}}{\longrightarrow} \Gamma^p(V) \longrightarrow V^{(1)} \longrightarrow 0,$$

where $\Phi(v) = v^p \in \mathrm{S}^p(V)$, while $\Gamma^p(V) \to V$ is the dual of $\Phi$, and it takes $v^{[p]}$ to $v$. To see this, one uses the exponential property of symmetric and divided powers to reduce to the case, when $\dim(V) = 1$. In this case $\mathrm{S}^p(V)$ and $\Gamma^p(V)$ are spanned by $v^p$ and $v^{[p]}$ respectively, where $v$ is a generator of $V$. Since $\mathrm{N}(v^p) = p! v^{[p]} = 0$, exactness of the sequence is obvious.

Let $\mathbb{K} \to \mathbb{L}$ be a field extension. The symmetric, exterior and divided powers have a nice behavior with respect to extension of scalars. For example, for any $\mathbb{K}$-vector space $V$, there are isomorphisms

$$\mathrm{S}^*_{\mathbb{L}}(V \otimes_{\mathbb{K}} \mathbb{L}) \cong (\mathrm{S}^*_{\mathbb{K}}(V)) \otimes_{\mathbb{K}} \mathbb{L},$$

Here the subscript $\mathbb{K}$ or $\mathbb{L}$ indicates the field, over which the tensors are taken. In particular, one can apply the isomorphisms to the Frobenius $\Phi : \mathbb{K} \to \mathbb{K}$ to get

$$(\mathrm{S}^*(V))^{(1)} \cong \mathrm{S}^*(V^{(1)}).$$

Similarly for divided and exterior powers.

**1.6. The Koszul complex.** — Let $A$ be a commutative $\mathbb{K}$-algebra and let $E$ be a finitely generated free $A$-module of rank $n$. Let $f : E \to A$ be a homomorphism of $A$-modules. We let $J$ be the image of $f$, so $J$ is an ideal of $A$. The *Koszul complex associated to* $f$ is the following chain complex $\mathrm{Kos}_*(f)$, where $\mathrm{Kos}_m(f) = \Lambda^m_A E$ is the

$m$-th exterior power of $E$ over the ring $A$, while the boundary map $\kappa_m : \Lambda_A^m E \to \Lambda_A^{m-1} E$ is given by

$$\kappa_m(x_1 \wedge \cdots \wedge x_m) = \sum_{i=1}^{m} (-1)^{i-1} f(x_i) x_1 \wedge \cdots \wedge x_{i-1} \wedge x_{i+1} \wedge \cdots \wedge x_m.$$

Thus the Koszul complex looks as follows

$$\cdots \longrightarrow 0 \longrightarrow \Lambda_A^n E \longrightarrow \Lambda_A^{n-1} E \longrightarrow \cdots \longrightarrow \Lambda_A^2 E \longrightarrow E \xrightarrow{f} A.$$

It is obvious that $\mathrm{H}_0(\mathrm{Kos}_*(f)) \cong A/J$. For a vector space $V$, we take $A = \mathrm{S}^*(V)$, $E = V \otimes A$ and define $f : E \to A$ to be the unique $A$-linear map, such that

$$f(v \otimes 1) = v \in V \subset \mathrm{S}^*(V).$$

In this case $\Lambda_A^k(E) \cong \Lambda^k(V) \otimes \mathrm{S}^*(V)$. Since $\mathrm{S}^*(V)$ is graded, the vector space $\Lambda^*(V) \otimes \mathrm{S}^*(V)$ is bigraded. One observes that $\kappa$ is of bidegree $(-1, 1)$, that is

$$\kappa_m(\Lambda^m(V) \otimes \mathrm{S}^j(V)) \subset \Lambda^{m-1}(V) \otimes \mathrm{S}^{j+1}(V).$$

Thus the Koszul complex splits into the direct sum of the complexes

$$0 \longrightarrow \Lambda^m(V) \longrightarrow \Lambda^{m-1}(V) \otimes V \longrightarrow \cdots \longrightarrow \Lambda^i(V) \otimes \mathrm{S}^{m-i}(V) \longrightarrow \cdots \longrightarrow \mathrm{S}^m(V).$$

**Theorem 1.1**. — *For any $m \geqslant 1$ there is an exact sequence:*

$$0 \longrightarrow \Lambda^m(V) \longrightarrow \cdots \longrightarrow \Lambda^i(V) \otimes \mathrm{S}^{m-i}(V) \longrightarrow \cdots \longrightarrow \mathrm{S}^m(V) \longrightarrow 0,$$

*where $\kappa_i : \Lambda^i(V) \otimes \mathrm{S}^{m-i}(V) \to \Lambda^{i-1}(V) \otimes \mathrm{S}^{m-i+1}(V)$ is given by*

$$\kappa_i(x_1 \wedge \cdots \wedge x_i \otimes y_1 \cdots y_{m-i}) = \sum_{j-1}^{i} (-1)^{j-1} x_1 \wedge \cdots \wedge x_{j-1} \wedge x_{j+1} \wedge \cdots \wedge x_i \otimes x_j y_1 \cdots y_{m-i}.$$

*Proof*. — Since

$$\Lambda^*(V \oplus W) \otimes \mathrm{S}^*(V \oplus W) \cong (\Lambda^*(V) \otimes \mathrm{S}^*(V)) \otimes (\Lambda^*(W) \otimes \mathrm{S}^*(W)),$$

it follows from the Künneth theorem that it suffices to consider the case when $\dim(V) = 1$. In this case the statement is obvious. □

It follows from the duality between the symmetric and divided powers that there is also an exact sequence:

$$0 \longrightarrow \Gamma^n(V) \longrightarrow \cdots \longrightarrow \Lambda^{n-i}(V) \otimes \Gamma^i(V) \longrightarrow \cdots \longrightarrow \Lambda^n(V) \longrightarrow 0.$$

**1.7. The de Rham complex and the Cartier isomorphism.** — Let $A$ be a commutative $\mathbb{K}$-algebra and $M$ be an $A$-module. A *derivation* from $A$ to $M$ is a linear map $D : A \to M$ satisfying the identity

$$D(ab) = aD(b) + bD(a).$$

The vector space of all derivations from $A$ to $M$ is denoted by $\mathrm{Der}(A, M)$. The module of Kähler differentials $\Omega^1_{A/\mathbb{K}}$, or simply $\Omega^1_A$, is the $A$-module generated by the symbols $dx$, $x \in A$ modulo the relations

$$d(x + y) = d(x) + d(y), \quad d(\lambda x) = \lambda d(x), \quad d(xy) = xd(y) + yd(x).$$

Here $x, y \in A$ and $\lambda \in \mathbb{K}$. Thus for any $A$-module $M$ there is an isomorphism

$$\mathrm{Hom}_A(\Omega^1_A, M) \cong \mathrm{Der}(A, M).$$

For any commutative algebra $A$ we let $\Omega^n_A$ be the $n$-th exterior power (over $A$) of the $A$-module $\Omega^1_A$. There is a well-defined linear map $d : \Omega^n_A \to \Omega^{n+1}_A$ given by

$$d(adb_1 \wedge \cdots \wedge db_n) = da \wedge db_1 \wedge \cdots \wedge da_n.$$

One has

$$d(\alpha \wedge \beta) = d(\alpha) \wedge \beta + (-1)^n \alpha \wedge d(\beta),$$

where $\alpha \in \Omega^n_A, \beta \in \Omega^m_A$ and $d^2 = 0$. Thus $\Omega^*_A$ is a commutative graded differential algebra, known as *de Rham complex*, whose cohomology is denoted by $\mathrm{H}^*_{DR}(A)$.

Let $A$ be a commutative algebra over a field $\mathbb{K}$ of characteristic $p > 0$. It is easy to check that the coboundary map in the de Rham complex $\Omega^*_A$ is $A^{(1)}$-linear. Here as usual $A^{(1)}$ is the Frobenius twist of $A$ and the $A^{(1)}$-module structure on $\Omega^*_A$ is given by:

$$a^{(1)} \star \omega = a^p \omega.$$

Thus $\mathrm{H}^*_{DR}(A)$ has the natural $A^{(1)}$-module structure, given by the same formula. Let $\mathrm{C}^{-1} : A \to \mathrm{H}^1_{DR}(A)$ be the map which assigns the class of $a^{p-1}da$ in $\mathrm{H}^1_{DR}(A)$ to $a \in A$. We have

$$\mathrm{C}^{-1}(a + b) = \mathrm{C}^{-1}(a) + \mathrm{C}^{-1}(b),$$

because $d\Theta(a, b) = (a + b)^{p-1}d(a + b) - a^{p-1}da - b^{p-1}db$, where $\Theta(x, y) \in \mathbb{Z}[x, y]$ is the polynomial with integral coefficients given by $p\Theta(x, y) = (x + y)^p - x^p - y^p$. Since $\mathrm{C}^{-1}(ab) = a \star \mathrm{C}^{-1}(b) + b \star \mathrm{C}^{-1}(a)$, there is a unique algebra homomorphism (still denoted by $\mathrm{C}^{-1}$)

$$\mathrm{C}^{-1} : \Omega^*_{A^{(1)}} \longrightarrow \mathrm{H}^*_{DR}(A),$$

which is given by $a \mapsto a^p$ in degree zero and which assigns $a^p b^{p-1}db$ to $a^{(1)}db^{(1)}$.

If $A = \mathrm{S}^*(V)$ is the symmetric algebra we have $\mathrm{Der}(\mathrm{S}^*(V), M) \cong \mathrm{Hom}(V, M)$ and therefore $\Omega^1_{\mathrm{S}^*(V)} \cong V \otimes \mathrm{S}^*(V)$. This follows from the fact that any derivation $D : \mathrm{S}^*(V) \to M$ is uniquely determined by its values on generators, because $\mathrm{S}^*(V)$

is a free commutative algebra. Therefore we have $\Omega^m_{\mathrm{S}^*(V)} \cong \Lambda^m(V) \otimes \mathrm{S}^*(V)$ and the coboundary map $d$ is given by

$$d(x_1 \wedge \cdots \wedge x_m \otimes y_1 \cdots y_k) = \sum_{i=1}^{k} (-1)^m x_1 \wedge \cdots \wedge x_m \wedge y_i \otimes y_1 \cdots y_{i-1} y_{i+1} \cdots y_k.$$

Thus $d(\Lambda^m(V) \otimes \mathrm{S}^k(V)) \subset \Lambda^{m+1}(V) \otimes \mathrm{S}^{k-1}(V)$ and the de Rham complex for the symmetric algebra splits as a direct sum of the following complexes

$$\Omega^*_m(V) := (\cdots \longleftarrow 0 \longleftarrow \Lambda^m(V) \longleftarrow \cdots \longleftarrow \Lambda^i(V) \otimes \mathrm{S}^{m-i}(V) \longleftarrow \cdots \longleftarrow \mathrm{S}^m(V)).$$

First consider the case $\dim(V) = 1$. In this case $A = \mathbb{K}[t]$ and the de Rham complex of $A$ is zero in dimensions $> 1$, equals $\mathbb{K}[t]$ in dimension zero and it is a free $\mathbb{K}[t]$-module generated by $dt$ in dimension one. The coboundary map $d : \mathbb{K}[t] \to \mathbb{K}[t]dt$ is given by $d(x^k) = kx^{k-1}dx$. Therefore $\mathrm{H}^i_{DR}(\mathbb{K}[t]) = 0$, if $i > 0$ and $\mathrm{H}^0_{DR}(\mathbb{K}[t]) = \mathbb{K}$, provided $\mathrm{Char}(\mathbb{K}) = 0$. If $\mathrm{Char}(\mathbb{K}) = p > 0$, then $\mathrm{H}^0_{DR}(\mathbb{K}[t])$ is spanned by $1, t^p, t^{2p}, \ldots$ if $i = 0$, while $\mathrm{H}^1_{DR}(\mathbb{K}[t])$ is spanned by the classes of $t^{p-1}dt, t^{2p-1}dt, \ldots$. Thus in this case $\mathrm{C}^{-1} : \Omega^*_{A^{(1)}} \to \mathrm{H}^*_{DR}(A)$ is an isomorphism. Since $\Omega^*_{\mathrm{S}^*(V \oplus W)} \cong \Omega^*_{\mathrm{S}^*(V)} \otimes \Omega^*_{\mathrm{S}^*(W)}$ we can use the Künneth theorem to get the following

**Lemma 1.2**. — *If* $\mathrm{Char}(\mathbb{K}) = 0$ *then*

$$\mathrm{H}^i_{DR}(\mathrm{S}^*(V)) = 0, i > 0, \quad \text{and} \quad \mathrm{H}^0_{DR}(\mathrm{S}^*(V)) \cong \mathbb{K}.$$

*If* $\mathrm{Char}(\mathbb{K}) = p > 0$ *and* $A = \mathrm{S}^*(V)$, *then* $\mathrm{C}^{-1} : \Omega^*_{A^{(1)}} \to \mathrm{H}^*_{DR}(A)$ *is an isomorphism.*

The inverse of the isomorphism $\mathrm{C}^{-1}$ is denoted by $\mathrm{C} : \mathrm{H}^*_{DR}(\mathrm{S}^*(V)) \to \Omega^*_{\mathrm{S}^*(V^{(1)})}$ and is called the *Cartier homomorphism* [**4**].

One easily shows that the boundary $\kappa : \Lambda^i \otimes \mathrm{S}^{n-i} \to \Lambda^{i-1} \otimes \mathrm{S}^{n-i+1}$ in the Koszul complex and the coboundary $d : \Lambda^{i-1} \otimes \mathrm{S}^{n-i+1} \to \Lambda^i \otimes \mathrm{S}^{n-i}$ in the de Rham complex are related by the following relation (the Euler formula):

$$(6) \qquad\qquad d\kappa + \kappa d = n \, \mathrm{Id}_{\Omega^i_n} : \Lambda^i \otimes \mathrm{S}^{n-i} \longrightarrow \Lambda^i \otimes \mathrm{S}^{n-i}.$$

**1.8. Adjoint functors and Ext.** — Let $\mathcal{C}$ and $\mathcal{D}$ be categories and let $F : \mathcal{C} \to \mathcal{D}$ and $G : \mathcal{D} \to \mathcal{C}$ be functors. We will say that $F$ is left adjoint to $G$ (or $G$ is a right adjoint of $F$) if for any $A \in \mathcal{C}$ and $B \in \mathcal{D}$ there is given an isomorphism

$$(7) \qquad\qquad \mathrm{Hom}_{\mathcal{D}}(F(A), B) \cong \mathrm{Hom}_{\mathcal{C}}(A, G(B))$$

which is natural with respect to $A$ and $B$. In this case we will say that $(F, G)$ is an adjoint pair from $\mathcal{C}$ to $\mathcal{D}$. Note that a functor right adjoint to $F$ is unique up to isomorphism. By taking $B = F(A)$ and looking at the image of $1_{F(A)}$ under the isomorphism (7) one obtains a natural transformation $u : \mathrm{Id}_{\mathcal{C}} \to G \circ F$, which is called *the unit of adjunction*. Putting $A = G(B)$, one gets a *counit of adjunction* $v : F \circ G \to 1_{\mathcal{D}}$. These transformations satisfy the following relations

$$(8) \qquad\qquad (vF) \circ (Fu) = 1_F \quad \text{and} \quad (Gv) \circ (uG) = 1_G.$$

Conversely, if two functors $F : \mathcal{C} \to \mathcal{D}$ and $G : \mathcal{D} \to \mathcal{C}$ and natural transformations $u : 1_\mathcal{C} \to G \circ F$ and $v : F \circ G \to 1_\mathcal{D}$ are given and $(vF) \circ (Fu) = 1_F$ and $(Gv) \circ (uG) = 1_G$ hold, then $(F, G)$ is an adjoint pair and the corresponding isomorphism $\mathrm{Hom}_\mathcal{D}(F(A), B) \to \mathrm{Hom}_\mathcal{C}(A, G(B))$ assigns $G(f) \circ u_A$ to $f : F(A) \to B$.

For a small category $\mathcal{C}$ and a category $\mathcal{E}$ we let $\mathrm{Func}(\mathcal{C}, \mathcal{E})$ be the category of all functors from $\mathcal{C}$ to $\mathcal{E}$. Then any functor $F : \mathcal{C} \to \mathcal{D}$ between small categories yields a functor $F^* : \mathrm{Func}(\mathcal{D}, \mathcal{E}) \to \mathrm{Func}(\mathcal{C}, \mathcal{E})$ by $R \mapsto R \circ F$. Here $R : \mathcal{D} \to \mathcal{E}$ is a functor.

**Lemma 1.3**. — *Let $(F, G)$ be an adjoint pair from $\mathcal{C}$ to $\mathcal{D}$. Assume $\mathcal{C}$ and $\mathcal{D}$ are small categories. Then for any category $\mathcal{E}$ the pair $(G^*, F^*)$ is a pair of adjoint functors from $\mathrm{Func}(\mathcal{C}, \mathcal{E})$ to $\mathrm{Func}(\mathcal{D}, \mathcal{E})$.*

*Proof.* — Take any functor $R : \mathcal{C} \to \mathcal{E}$ and apply it to $u(x) : x \to GF(x)$, $x \in C$ to get the transformation $R(u) : R \to R(GF)$. Varying $R$, we get the transformation $u^* : 1_{\mathrm{Func}(\mathcal{C}, \mathcal{E})} \to F^*G^*$. Similarly, one gets $v^* : G^*F^* \to 1_{\mathrm{Func}(\mathcal{D}, \mathcal{E})}$. It is clear that the transformations $(u^*, v^*)$ satisfy the relations (8) and hence the Lemma. □

**Lemma 1.4**. — *Let $\mathbf{A}$ and $\mathbf{B}$ be abelian categories with enough projective and injective objects. Moreover assume $F : \mathbf{A} \to \mathbf{B}$ is a functor which has a right adjoint functor $G : \mathbf{B} \to \mathbf{A}$. Then the following conditions are equivalent:*

(i) *$F$ is exact and preserves projective objects*
(ii) *$G$ is exact and preserves injective objects*
(iii) *For all $A \in \mathbf{A}$ and $B \in \mathbf{B}$ there is an isomorphism*

$$\mathrm{Ext}_\mathbf{B}^*(F(A), B) \cong \mathrm{Ext}_\mathbf{A}^*(A, G(B)).$$

(iv) *$F$ and $G$ are exact functors.*
(v) *For all $A \in \mathbf{A}$ and $B \in \mathbf{B}$ there is an isomorphism*

$$\mathrm{Ext}_\mathbf{B}^1(F(A), B) \cong \mathrm{Ext}_\mathbf{A}^1(A, G(B)).$$

*Proof.* — It is clear that (iii) $\Rightarrow$ (v). We first prove that (i) $\Rightarrow$ (iii). Let $P_* \to A$ be a projective resolution. By assumption $F(P_*) \to F(A)$ is also a projective resolution. Therefore we have

$$\mathrm{Ext}_\mathbf{B}^*(F(A), B) = \mathrm{H}^*(\mathrm{Hom}_\mathbf{B}(F(P_*), B)) = \mathrm{H}^*(\mathrm{Hom}_\mathbf{A}(P_*, G(B))) = \mathrm{Ext}_\mathbf{B}^*(A, G(B)).$$

The dual argument shows that (ii) $\Rightarrow$ (iii). Assume now that the condition (v) holds and let $P$ be a projective object in $\mathbf{A}$. Then

$$\mathrm{Ext}_\mathbf{B}^1(F(P), B) = \mathrm{Ext}_\mathbf{A}^1(P, G(B)) = 0$$

for all $B \in \mathbf{B}$ and hence $F(P)$ is projective. Similarly one shows that $G$ respects injective objects. Now we show that $F$ is an exact functor. To this end we consider an exact sequence

$$0 \longrightarrow A_1 \longrightarrow A \longrightarrow A_2 \longrightarrow 0$$

in $\mathbf{A}$. Since $F$ has a right adjoint it is right exact. We let $B$ be the kernel of $F(A_1) \to F(A_2)$. Thus we have an exact sequence

$$0 \longrightarrow B \longrightarrow F(A_1) \longrightarrow F(A) \longrightarrow F(A_2) \longrightarrow 0.$$

Now take an injective object $I \in \mathbf{B}$. Since $\mathrm{Hom}_{\mathbf{B}}(-, I)$ is an exact functor we have an exact sequence

$$0 \longrightarrow \mathrm{Hom}_{\mathbf{B}}(F(A_2), I) \longrightarrow \mathrm{Hom}_{\mathbf{B}}(F(A), I) \longrightarrow \mathrm{Hom}_{\mathbf{B}}(F(A_1), I)$$
$$\longrightarrow \mathrm{Hom}_{\mathbf{B}}(B, I) \longrightarrow 0.$$

By assumption $G$ is right adjoint to $F$. Thus we can rewrite

$$0 \longrightarrow \mathrm{Hom}_{\mathbf{A}}(A_2, G(I)) \longrightarrow \mathrm{Hom}_{\mathbf{A}}(A, G(I)) \longrightarrow \mathrm{Hom}_{\mathbf{A}}(A_1, G(I))$$
$$\longrightarrow \mathrm{Hom}_{\mathbf{B}}(B, I) \longrightarrow 0.$$

Since $G(I)$ is an injective object, the map $\mathrm{Hom}_{\mathbf{A}}(A, G(I)) \to \mathrm{Hom}_{\mathbf{A}}(A_1, G(I))$ is surjective and therefore $\mathrm{Hom}_{\mathbf{B}}(B, I) = 0$. Since $I$ was an arbitrary injective object and $\mathbf{B}$ has enough injective objects it follows that $B = 0$. Thus $F$ is an exact functor. The dual argument shows that $G$ is exact too. Therefore we proved (v)$\Rightarrow$ (i), (ii), (iv) and we have only to show (iv)$\Rightarrow$ (i). Let $P$ be a projective object in $\mathbf{A}$, then $\mathrm{Hom}_{\mathbf{B}}(F(P), -) \cong \mathrm{Hom}_{\mathbf{A}}(P, -) \circ G$ is a composite of two exact functors and therefore is exact. Hence $F(P)$ is a projective object. $\qquad\square$

**Lemma 1.5**. — *Let $(l, r)$ be an adjoint pair from $\mathbf{C}$ to $\mathbf{D}$. Assume $\mathbf{C}$ and $\mathbf{D}$ are small categories. Then for any $F : \mathbf{C} \to \mathcal{V}$ and $G : \mathbf{D} \to \mathcal{V}$ there is an isomorphism*

$$\mathrm{Ext}^*_{\mathrm{Func}(\mathbf{D}, \mathcal{V})}(F \circ r, G) \cong \mathrm{Ext}^*_{\mathrm{Func}(\mathbf{C}, \mathcal{V})}(F, G \circ l).$$

*Proof.* — This follows directly from Lemma 1.4, because the functor

$$(-) \circ r : \mathrm{Func}(\mathbf{C}, \mathcal{V}) \longrightarrow \mathrm{Func}(\mathbf{D}, \mathcal{V})$$

is left adjoint to the functor $(-) \circ l : \mathrm{Func}(\mathbf{D}, \mathcal{V}) \to \mathrm{Func}(\mathbf{C}, \mathcal{V})$ (see Lemma 1.3) and both of them are exact. $\qquad\square$

## 2. The category $\mathcal{F}(\mathbb{K})$

**2.1. Definitions and examples.** — We let $\mathcal{V}^f$ be the full subcategory of $\mathcal{V}$ consisting of finite dimensional vector spaces, furthermore we let $\mathcal{F}(\mathbb{K})$ or simply $\mathcal{F}$ be the category of all covariant functors $\mathcal{V}^f \to \mathcal{V}$.

**Example 2.1**. — We have the inclusion functor $\mathrm{Id} \in \mathcal{F}$ given by $\mathrm{Id}(V) = V$, $V \in \mathcal{V}^f$. The tensor, symmetric, exterior and divided powers define the functors

$$\mathrm{T}^n, \mathrm{S}^n, \Lambda^n, \Gamma^n \in \mathcal{F}.$$

For $n = 1$ we have $S^1 = \Lambda^1 = \Gamma^1 = \mathrm{Id}$. For any $V \in \mathcal{V}^f$ one defines $P_V \in \mathcal{F}$ and $I_V$ by

$$P_V(W) : = \mathbb{K}[\mathrm{Hom}(V, W)], \quad I_V(W) = \mathrm{Maps}(V^\#, W)$$

Here $\mathbb{K}[S]$ denotes the free vector space generated by a set $S$. If $V = \mathbb{K}$, then we write $P$ and $I$ instead of $P_V$ and $I_V$.

If $\mathbb{K}$ is a field of characteristic $p > 0$, then the *Frobenius twist* is the endofunctor

$$(-)^{(1)} : \mathcal{F} \longrightarrow \mathcal{F}, \quad F \longmapsto F^{(1)}$$

defined by $F^{(1)}(V) := F(V^{(1)})$. By induction one defines $F^{(r+1)} = (F^{(r)})^{(1)}, r \geqslant 0$. For example $\mathrm{Id}^{(1)}(V) = V^{(1)}$. One has a natural transformation $S^{n(1)} \to S^{np}$ given by $v^{(1)} \mapsto v^p$. Let us note that for the finite prime field $\mathbb{F}_p$ the functors $\mathrm{Id}^{(1)}$ and Id are isomorphic in $\mathcal{F}$ and the Frobenius twist induces an equivalence of categories $(-)^{(1)} : \mathcal{F} \to \mathcal{F}$.

The category $\mathcal{F}$ as any category of functors into the category of vector spaces is an abelian category with enough projective and injective objects (see [**7**, Section 2.3] in this volume). We just recall that kernels, cokernels, products and coproducts in $\mathcal{F}$ are calculated pointwise. In other words, if $\xi : F \to G$ is a morphism in $\mathcal{F}$, then $\mathrm{Ker}(\xi)$ and $\mathrm{Coker}(\xi)$ are objects in $\mathcal{F}$ given by $V \mapsto \mathrm{Ker}(\xi_V)$ and $V \mapsto \mathrm{Coker}(\xi_V)$ respectively. Projective generators in $\mathcal{F}$ are $P_V$, $V \in \mathcal{V}^f$, while injective cogenerators are $I_V$, $V \in \mathcal{V}^f$ (compare [**7**, Section 2.3]). These facts are trivial consequences of the Yoneda lemma, which says that $\mathrm{Hom}_{\mathcal{F}}(P_V, F) \cong F(V)$, and the isomorphism (9) below.

We also need the pointwise tensor product in $\mathcal{F}$, which is defined as follows: if $F \in \mathcal{F}$ and $G \in \mathcal{F}$ then $F \otimes G \in \mathcal{F}$ is given by

$$(F \otimes G)(V) := F(V) \otimes G(V).$$

For example one has

$$P_V \otimes P_W \cong P_{V \oplus W}, \quad I_V \otimes I_W \cong I_{V \oplus W}, \quad T^n \cong \mathrm{Id}^{\otimes n}.$$

For an object $F \in \mathcal{F}$ one defines the *dual* $F^\sharp$ of $F$ by

$$F^\sharp(V) = (F(V^\#))^\#.$$

For example, we have $(S^n)^\sharp = \Gamma^n$, $(\Lambda^n)^\sharp \cong \Lambda^n$, $(T^n)^\sharp \cong T^n$, $(P_V)^\sharp \cong I_V$. In general $F^{\sharp\sharp} \cong F$ provided $F(V)$ is finite dimensional for all $V \in \mathcal{V}^f$. For any $F, G \in \mathcal{F}$ there is a natural isomorphism

(9)                          $$\mathrm{Hom}_{\mathcal{F}}(F, G^\sharp) \cong \mathrm{Hom}_{\mathcal{F}}(G, F^\sharp).$$

If $F \in \mathcal{F}$ then there is a canonical way to prolong $F$ as a functor $\mathcal{V} \to \mathcal{V}$, by

$$F(U) := \mathrm{colim}_{V \in \mathrm{s}(U)} F(V),$$

where $\mathrm{s}(U)$ is the poset of finite dimensional subspaces of a vector space $U$. In this way one obtains an equivalence between the category $\mathcal{F}$ and the category of endofunctors of $\mathcal{V}$ which commute with filtered colimits. Since the category of such endofunctors is closed under composition, this equivalence of categories allows us to consider the composition as a bifunctor $\circ : \mathcal{F} \times \mathcal{F} \to \mathcal{F}$. The bifunctor $\circ$ is additive with respect to the first variable, but NOT with respect to the second variable.

**2.2. Polynomial functors *à la* Eilenberg and MacLane.** — Let $F$ be a functor $F : \mathcal{V}^f \to \mathcal{V}$. For each $V \in \mathcal{V}^f$ one can consider the canonical retraction $r_V : V \oplus \mathbb{K} \to V$. Then one defines the functor $\Delta F : \mathcal{V}^f \to \mathcal{V}$ by

$$\Delta F(V) := \mathrm{Ker}(F(r_V) : F(V \oplus \mathbb{K}) \longrightarrow F(V)).$$

In this way one obtains an endofunctor $\Delta : \mathcal{F} \to \mathcal{F}$. We let $\Delta^k$ be the $k$-th iteration of $\Delta$. Since $r_V$ has a canonical section, we have $F(V \oplus K) = F(V) \oplus \Delta F(V)$.

A functor $F$ is a *polynomial functor* in the sense of Eilenberg and MacLane if $\Delta^k F = 0$ for some $k \geqslant 0$. If $\Delta^{n+1} F = 0$ but $\Delta^n F \neq 0$, then we say that $F$ is *of degree $n$*.

In this case we write $\deg(F) = n$. It is clear that the degree zero functors are just constant functors. We claim that the definition of the degree of functors coincides with the one given in [7, Section 4.1] *via* cross-effects, in other words $\Delta^k F = 0$ if and only if $\mathrm{Cr}_k F = 0$. Indeed, comparing the definitions one sees that

$$(10) \qquad \Delta^k F(V) \cong \mathrm{Cr}_k F(\mathbb{K}, \dots, \mathbb{K}) \oplus \mathrm{Cr}_{k+1} F(V, \mathbb{K}, \dots, \mathbb{K})$$

Thus, if $\mathrm{Cr}_k F = 0$, then $\Delta^k F = 0$ as well. Conversely, if $\Delta^k F = 0$, then $\mathrm{Cr}_k F(\mathbb{K}, \dots, \mathbb{K}) = 0$ and $\mathrm{Cr}_{k+1} F(V, \mathbb{K}, \dots, \mathbb{K}) = 0$. The second isomorphism shows that $\mathrm{Cr}_k F(V, \mathbb{K}, \dots, \mathbb{K})$ is an additive functor in $V$. Since the cross-effect is symmetric in its variables, we see that $\mathrm{Cr}_k F$ is a $k$-linear functor defined on $\mathcal{V}^f$. Thus $\mathrm{Cr}_k F(V_1, \dots, V_k)$ is a sum of $\prod \dim(V_i)$ copies of $\mathrm{Cr}_k F(\mathbb{K}, \dots, \mathbb{K})$, which is zero by assumption - hence the claim.

***Example 2.2***. — $\deg(F) = 0$ if and only if $F$ is constant and $\deg(F) = 1$ if and only if $F$ is of the form $F(V) = (A \otimes_{\mathbb{K}} V) \bigoplus C$ for some $C \in \mathcal{V}$ and a nontrivial $\mathbb{K}$-$\mathbb{K}$-bimodule $A$. One has

$$\deg(\Lambda^n) = \deg(\Gamma^n) = \deg(\mathrm{T}^n) = \deg(\mathrm{S}^n) = n.$$

We have

$$\mathrm{P}(V \oplus \mathbb{K}) = \mathbb{K}[V \oplus \mathbb{K}] \cong \mathbb{K}[V] \otimes \mathbb{K}[\mathbb{K}] = \mathrm{P}(V) \otimes \mathbb{K}[\mathbb{K}].$$

Hence P is a direct summand of $\Delta \mathrm{P}$ and thus P is not a polynomial functor. If $\mathrm{Char}(\mathbb{K}) = p > 0$, then $\deg \mathrm{I}^{(1)} = 1$.

***Lemma 2.3***. — *The functor $\Delta : \mathcal{F} \to \mathcal{F}$ is an exact functor and takes projective objects to projective objects.*

*Proof*. — The computation in the previous example shows that $\Delta \mathrm{P}_V$ is of the form $\mathrm{P}_V \otimes U$ with some vector space $U$. Since the coproduct of projective objects is also projective, it follows that $\mathrm{P}_V \otimes U$ is a projective object for any vector space $U$. $\square$

We let $\mathcal{F}_d$ be the full subcategory of $\mathcal{F}$ consisting of functors of degree $\leqslant d$. The category $\mathcal{F}_d$ is closed under taking subobjects, extensions, quotients and (co)products. In particular $\mathcal{F}_d$ is an abelian category and the inclusion $\mathcal{F}_d \hookrightarrow F$ is an exact functor. This functor has both left and right adjoint functors and the category $\mathcal{F}_d$ has enough projective and injective objects (see [**7**, Lemma 3.3]). The tensor product yields a bifunctor $- \otimes - : \mathcal{F}_n \otimes \mathcal{F}_m \to \mathcal{F}_{n+m}$.

A functor $F$ is called *analytic* if it is a union of subfunctors of finite degrees. The subcategory $\mathcal{F}_\omega$ of analytic functors is also closed under subobjects, extensions, quotients and (co)products. It is also an abelian category and the inclusion functor $\mathcal{F}_\omega \hookrightarrow \mathcal{F}$ is an exact functor, which has the right adjoint functor, which is given by $F \mapsto \bigcup_{n\geqslant 0} \mathrm{t}^n F$, where $\mathrm{t}^n$ is defined in [**7**, Lemma 2.13]. It follows, that the category of analytic functors $\mathcal{F}_\omega$ has enough injective objects.

The functors $\mathrm{T}^* = \bigoplus_{d\geqslant 0} \mathrm{T}^d$, $\Lambda^*$, $\Gamma^*$ are examples of analytic functors. We leave as an exercise to show that P is not an analytic functor. The injective functor I is an analytic functor provided $\mathbb{K}$ is a finite field (see Lemma 2.5) below.

**Lemma 2.4**. — *Let $K = \mathbb{F}_q$ be a finite field with $q = p^s$ elements. For any finite dimensional vector space $V$ there is a natural isomorphism*

$$\mathrm{I}(V) \cong \mathrm{S}^*(V)/(v^q - v),$$

*where the right hand side is the quotient of the symmetric algebra generated by $V$ by the ideal generated by the elements $v^q - v$, $v \in V$.*

*Proof.* — Since $\mathrm{I}(V) \cong \mathrm{Maps}(V^\#, \mathbb{K})$, it has a natural commutative $\mathbb{K}$-algebra structure. Moreover the equality $a^q = a$ holds in $\mathbb{K}$ and therefore in $\mathrm{I}(V)$ as well. Hence the natural embedding $V \subset \mathrm{Maps}(V^\#, \mathbb{K})$ yields a homomorphism of algebras $\mathrm{S}^*(V)/(v^q - v) \to \mathrm{I}(V)$. Surjectivity of this map follows from the fact that any map between finite vector spaces is given by a polynomial, while injectivity follows from a counting of dimensions. $\square$

**Lemma 2.5**. — *Over finite fields the functor $\mathrm{I}_V$ is analytic for all $V \in \mathcal{V}^f$.*

*Proof.* — Analyticity of I is a consequence of the previous lemma, because a quotient of an analytic functor is analytic and $\mathrm{S}^*(V)$ is analytic. The general case follows from the isomorphism $\mathrm{I}_V \cong \mathrm{I}^{\otimes(\dim V)}$ and the fact that tensor product of two analytic functors is analytic too. $\square$

**2.3. Type** $(\mathrm{FP})_\infty$. — A functor $F \in \mathcal{F}$ is *finitely generated* if there is an epimorphism $\mathrm{P}_{V_1} \oplus \cdots \oplus \mathrm{P}_{V_k} \to F$, for some $V_1, \ldots, V_k \in \mathcal{V}^f$. It is clear that each direct sum of two finitely generated functors is finitely generated.

A functor $F$ is said to be of type $(\mathrm{FP})_\infty$ if there is a projective resolution $P_* \to F$ with $P_i$ finitely generated for all $i$ (compare with [**1**]).

A poset $I$ is called colim-*exact* if the functor colim : $\mathrm{Func}(I, \mathcal{V}) \to \mathcal{V}$ is exact. For example, discrete or filtered posets are colim-exact. The proof of the following result is completely analogous to [**1**, Theorem 1.3].

**Proposition 2.6**. — *The following conditions are equivalent for $F \in \mathcal{F}$:*

(i) *$F$ is of type* $(\mathrm{FP})_\infty$

(ii) *For any colim-exact poset $I$ and a functor $M : I \to \mathcal{F}$, $i \mapsto M_i$, the natural map* colim $\mathrm{Ext}_{\mathcal{F}}^k(F, M_i) \to \mathrm{Ext}_{\mathcal{F}}^k(F, \mathrm{colim}\, M_i)$ *is an isomorphism for all $k$.*

The following result is due to L. Schwartz [**6**, Proposition 10.1].

**Theorem 2.7 (L. Schwartz)**. — *Let $\mathbb{K}$ be a finite field. Assume $F \in \mathcal{F}$ has finite degree and takes values in the category of finite dimensional vector spaces. Then $F$ is of type* $(\mathrm{FP})_\infty$.

Before giving the proof we need some additional notation. A functor $F$ is *generated in dimension $n$*, if the canonical map

$$\mathrm{ev}^{V,F} : F(V) \otimes \mathrm{P}_V \longrightarrow F$$

given by

$$\mathrm{ev}_W^{V,F}(a \otimes f) = f_*(a), \quad a \in F(V), \ f : V \longrightarrow W$$

is an epimorphism, provided $\dim V = n$. Let us observe that any finitely generated functor is generated in dimension $n$ for some $n \geqslant 0$. This follows from the fact that, that for any $V, W \in \mathcal{V}^f$ the projections $V \oplus W \to V$ and $V \oplus W \to W$ yield an epimorphism $\mathbb{K} \oplus \mathrm{P}_{V \oplus W} \to \mathrm{P}_V \oplus \mathrm{P}_W$, where $\mathbb{K}$ is considered as a constant functor.

**Lemma 2.8**. — *Let $\mathbb{K}$ be a finite field and $F \in \mathcal{F}$. Then*

(i) *$F$ is finitely generated if and only if $F(V) \in \mathcal{V}^f$ for all $V \in \mathcal{V}^f$ and $F$ is generated in dimension $n$ for some $n$.*

(ii) *Assume $\Delta^k F$ is generated in dimension $n$, then $F$ is generated in dimension $n + k$.*

*Proof*

(i) The functor $\mathrm{P}_V$ has values in $\mathcal{V}^f$, therefore the same is true for any finitely generated $F \in \mathcal{F}$. Conversely, assume $F(V) \in \mathcal{V}^f$ provided $V \in \mathcal{V}^f$. Then $F(V) \otimes \mathrm{P}_V$ is finitely generated, and the same can be said about $F$, provided $F$ is generated in dimension $n$.

(ii) It suffices to consider the case $k = 1$. We let $X \in \mathcal{F}$ be the cokernel of the canonical map

$$\mathrm{ev}^{V \oplus \mathbb{K}, F} : F(V \oplus \mathbb{K}) \otimes \mathrm{P}_{V \oplus \mathbb{K}} \longrightarrow F.$$

We have to prove $X = 0$. It suffices to show $\Delta X = 0$. Indeed, this condition implies that $X$ is constant. On the other hand

$$\mathrm{ev}_0^{V \oplus \mathbb{K}, F} : F(V \oplus \mathbb{K}) \otimes \mathrm{P}_{V \oplus \mathbb{K}}(0) \longrightarrow F(0)$$

is surjective and hence $X(0) = 0$ which shows that $X = 0$. In order to prove $\Delta X = 0$ we first observe that the endofunctor $\Delta : \mathcal{F} \to \mathcal{F}$ is an exact functor and preserves coproducts. Therefore $\Delta(U \otimes T) = U \otimes \Delta(T)$, for any vector space $U$. It follows that there is an exact sequence

$$F(V \oplus \mathbb{K}) \otimes \Delta \mathrm{P}_{V \oplus \mathbb{K}} \longrightarrow \Delta F \longrightarrow \Delta X \longrightarrow 0.$$

For a vector space $W$ we let $\pi_W$ be the projection $W \oplus \mathbb{K} \to W$. Then the formal difference

$$\delta_W = 1_{W \oplus \mathbb{K}} - \pi_W \in \mathbb{K}[\mathrm{Hom}_{\mathbb{K}}(W \oplus \mathbb{K}, W \oplus \mathbb{K})] = \mathrm{P}_{W \oplus \mathbb{K}}(W \oplus \mathbb{K})$$

lies in $\Delta \mathrm{P}_{W \oplus \mathbb{K}}(W)$. By the Yoneda lemma $\delta_W$ yields a natural transformation $\overline{\delta}_W :$ $\mathrm{P}_W \to \Delta \mathrm{P}_{W \oplus \mathbb{K}}$. We let $i$ to be the inclusion $\Delta F(V) \to F(V \oplus \mathbb{K})$ and $\overline{\delta} = \delta_V$. Now one checks that the following diagram is commutative:

$$
\begin{array}{c}
\Delta F(V) \otimes \mathrm{P}_V \\
{\scriptstyle i \otimes \overline{\delta}} \downarrow \qquad \searrow {\scriptstyle \mathrm{ev}^{V, \Delta F}} \\
F(V \oplus \mathbb{K}) \otimes \Delta \mathrm{P}_{V \oplus \mathbb{K}}(V) \longrightarrow \Delta F \longrightarrow \Delta X \longrightarrow 0.
\end{array}
$$

By assumption the homomorphism $\mathrm{ev}^{\Delta F, V} : \Delta F(V) \otimes \mathrm{P}_V \to \Delta F$ is a surjection. Therefore the map $F(V \oplus \mathbb{K}) \otimes \Delta \mathrm{P}_{V \oplus \mathbb{K}}(V) \to \Delta F$ is surjective and $\Delta X = 0$. $\square$

**Corollary 2.9**. — *Let $\mathbb{K}$ be a finite field. If $F \in \mathcal{F}$ has finite degree then $F$ is generated in dimension $m$, for some $m \geqslant 0$.*

*Proof*. — The statement is clear, because if $F$ is of finite degree then $\Delta^d F = 0$, where $d > \deg(F)$. $\square$

**Lemma 2.10**. — *Let $\mathbb{K}$ be a finite field and let $F : \mathcal{V}^f \to \mathcal{V}^f$ be a functor. Suppose $F$ is generated in dimension $n$ and $\Delta^k F$ is a projective functor for some $k > 0$. Then $F$ is of type $(\mathrm{FP})_\infty$.*

*Proof*. — Let $T$ be the kernel of the canonical map

$$\mathrm{ev}^{V, F} : F(V) \otimes \mathrm{P}_V \longrightarrow F,$$

where $\dim(V) = n$. By assumptions $\mathrm{ev}^{F, V}$ is surjective. On the other hand $F(V) \otimes \mathrm{P}_V$ is a projective object in $\mathcal{F}$. Since $\Delta^k$ is an exact functor, which takes projective objects to projective objects and since $\Delta^k F$ is projective $\Delta^k(\mathrm{ev}^{V, F})$ splits and $\Delta^k T$ is also a projective object which is generated in dimension $n$. By the previous result $T$ is generated in dimension $k + n$. Thus $T$ satisfies the assumption of the theorem (one needs to replace $n$ by $n + k$) and therefore one can iterate this process. $\square$

*Proof of Theorem 2.7*. — One observes that any polynomial functor $F$ is generated in dimension $n$ for some $n$ thanks to (ii) Lemma 2.8. Moreover $\Delta^k F = 0$ and therefore $\Delta^k F$ is projective provided $k > \deg(F)$. Now we are in the position to use Lemma 2.10. $\square$

**Remark 2.11.** — The following conjecture was posed by L. Schwartz.

*If $K$ is a finite field and $V \in \mathcal{V}^f$ then the injective object $\mathrm{I}_V$ is artinian.*

The conjecture implies the following assertions

*Any injective object in $\mathcal{F}_\omega$ is also injective in $\mathcal{F}$.*

*Any finitely generated functor is of type* $(\mathrm{FP})_\infty$.

Theorem 2.7 shows that one can prove the last assertion for some class of functors directly without using the conjecture.

## 2.4. Vanishing Lemma

**Lemma 2.12.** — *Let $A \in \mathcal{F}$ be an additive functor and let $B : \mathcal{V} \times \mathcal{V}^f \to \mathcal{V}$ be a bifunctor. We put $B^d(V) := B(V,V)$, $B^1(V) := B(V,0)$ and $B^2(V) := B(0,V)$, so that $B^d$, $B^1$, $B^2$ are in $\mathcal{F}$. There is an isomorphism:*

$$\mathrm{Ext}_{\mathcal{F}}^*(A, B^d) \cong \mathrm{Ext}_{\mathcal{F}}^*(A, B^1) \oplus \mathrm{Ext}_{\mathcal{F}}^*(A, B^2)$$

*Proof.* — We have functors $i_k \colon \mathcal{V}^f \to \mathcal{V}^f \times \mathcal{V}^f$, $k = 1,2,3$ given by: $i_1(V) = (V,0)$, $i_2(V) = (0,V)$, $i_3(V) = (V,V)$. We also have functors $p_k \colon \mathcal{V}^f \times \mathcal{V}^f \to \mathcal{V}^f$, $k = 1,2,3$ given by: $p_1(V,W) = V$, $p_2(V,W) = W$, $p_3(V,W) = V \oplus W$. For each $k = 1,2,3$, the pair $(i_k, p_k)$ is an adjoint pair. Thus Lemma 1.5 gives:

$$\mathrm{Ext}_{\mathcal{F}}^*(A, B^d) = \mathrm{Ext}_{\mathcal{F}}^*(A, B \circ i_3) \cong \mathrm{Ext}_{bi-\mathcal{F}}^*(A \circ p_3, B)$$

Here $bi - \mathcal{F}$ denotes the category of bifunctors $\mathcal{V}^f \times \mathcal{V}^f \to \mathcal{V}$. Since $A$ is an additive functor we have $A \circ p_3 = A \circ p_1 \oplus A \circ p_2$. Thus

$$\mathrm{Ext}_{bi-\mathcal{F}}^*(A \circ p_3, B) = \mathrm{Ext}_{bi-\mathcal{F}}^*(A \circ p_1, B) \oplus \mathrm{Ext}_{bi-\mathcal{F}}^*(A \circ p_2, B).$$

We can use Lemma 1.5 to obtain

$$\mathrm{Ext}_{bi-\mathcal{F}}^*(A \circ p_k, B) = \mathrm{Ext}_{\mathcal{F}}^*(A, B^k), \quad \text{for } k = 1,2$$

and we are done.                                                                    □

A functor $F$ from an additive category $\mathcal{A}$ to an additive category $\mathcal{B}$ is called *diagonalizable* if it is a composite of the form $F = T \circ i_3$, where as before $i_3 : \mathcal{A} \to \mathcal{A} \times \mathcal{A}$ is the diagonal map and $T : \mathcal{A} \times \mathcal{A} \to \mathcal{B}$ is a bifunctor satisfying $T(0, X) = 0 = T(X, 0)$ for every object $X$ in $\mathcal{A}$.

**Corollary 2.13** ([12]). — *If $F$ is a diagonalizable functor in $\mathcal{F}$ and $A$ is an additive functor, then*

$$\mathrm{Ext}_{\mathcal{F}}^*(A, F) = 0.$$

*In particular if $F'$ and $F''$ are two functors which map $0$ to $0$, then*

$$\mathrm{Ext}_{\mathcal{F}}^*(A, F' \otimes F'') = 0.$$

## 3. Computation of $\mathrm{Ext}^*_{\mathcal{F}}(\mathrm{Id}, \mathrm{S}^*)$

**3.1. The main theorem.** — In this section we take $\mathbb{K} = \mathbb{F}_p$ and we give the complete computation of $\mathrm{Ext}^*_{\mathcal{F}}(\mathrm{Id}, \mathrm{S}^*)$ following [**6**]. These groups previously were computed by L. Breen [**3**] using different methods.

***Theorem 3.1.*** — *One has*

$$\mathrm{Ext}^i_{\mathcal{F}}(\mathrm{Id}, \mathrm{S}^{p^h}) = \begin{cases} \mathbb{K} & \text{if } 2p^h \mid i, \\ 0 & \text{otherwise.} \end{cases}$$

***Remark 3.2.*** — If $m$ is not a power of $p$, then $\mathrm{Ext}^*_{\mathcal{F}}(\mathrm{Id}, \mathrm{S}^m) = 0$ thanks to Proposition 3.4 below. One can prove that there exists an isomorphism of graded associative algebras:

$$\mathrm{Ext}^*_{\mathcal{F}}(\mathrm{Id}, \mathrm{Id}) \cong \Gamma(t),$$

where the ring structure on $\mathrm{Ext}^*_{\mathcal{F}}(\mathrm{Id}, \mathrm{Id})$ is given by the Yoneda product. Let us recall that the divided power algebra $\Gamma(t)$ with one generator is isomorphic as an algebra to $\mathbb{F}_p[e_0, \ldots, e_h, \ldots]/(e_h{}^p; h \geqslant 0)$. Here $|e_h| = 2p^h$. Furthermore the class $e_0 \in \mathrm{Ext}^2_{\mathcal{F}}(\mathrm{Id}, \mathrm{Id})$ is given by the exact sequence (5). For the proof of these facts we refer the reader to [**6**].

The proof given below closely follows the original paper [**6**] (see also [**10**, Chap. 13]). In this section we write $\mathrm{H}^*(T)$ instead of $\mathrm{Ext}^*_{\mathcal{F}}(\mathrm{Id}, T)$. The key to the computation is Corollary 2.13, which says that $\mathrm{H}^*(T_1 \otimes T_2) = 0$ provided $T_1(0) = 0 = T_2(0)$ and the following lemma, known as Kuhn lemma (see [**6**]).

***Lemma 3.3.*** — *Let $F \in \mathcal{F}$ be a functor of type* $(\mathrm{FP})_\infty$. *Then for any $k > 0$ the colimit of the sequence*

$$\cdots \longrightarrow \mathrm{Ext}^k_{\mathcal{F}}(F, \mathrm{S}^{p^h}) \xrightarrow{\Phi_*} \mathrm{Ext}^k_{\mathcal{F}}(F, \mathrm{S}^{p^{h+1}}) \xrightarrow{\Phi_*} \cdots$$

*is zero, where $\Phi : \mathrm{S}^n \to \mathrm{S}^{pn}$ is given by the Frobenius in $\mathrm{S}^*(V)$.*

For the proof we refer to [**6**, Appendice] or [**14**] in this volume. Since $\mathrm{Id}$ is of type $(\mathrm{FP})_\infty$ thanks to Theorem 2.7, we have: $\mathrm{colim}_h \mathrm{H}^i(\mathrm{S}^{p^h}) = 0$ for all $i > 0$.

***Proposition 3.4.*** — *Assume $n \geqslant 2$.*
  (i) *If $n$ is not a power of $p$, then $\mathrm{H}^*(\mathrm{S}^n) = 0$.*
  (ii) *There is an isomorphism*

$$\mathrm{H}^i(\Lambda^n) \cong \mathrm{H}^{i-n+1}(\mathrm{S}^n).$$

*Proof*

  (i) By Corollary 2.13 we know that $\mathrm{H}^*(A, \mathrm{S}^i \otimes \mathrm{S}^j) = 0$, for $i, j > 1$. The composition $\mathrm{S}^n \to \mathrm{S}^i \otimes \mathrm{S}^{n-i} \to \mathrm{S}^n$ is the multiplication by the binomial coefficient $\binom{n}{i}$. Here the first (resp. second) map comes from the comultiplication (resp. multiplication) in the symmetric algebra. Thus $\mathrm{H}^*(\mathrm{S}^n)$ is annihilated by $\binom{n}{i}$, for $1 \leqslant i \leqslant n - 1$. Now it is enough to remark that the greatest common divisor of the numbers $\binom{n}{1}, \binom{n}{2}, \ldots, \binom{n}{n-1}$ is $p$, if $n$ is a power of $p$, and is 1, if $n$ is not a prime power.

(ii) Apply Corollary 2.13 to the exact sequence of Theorem 1.1. □

**3.2. Hypercohomology spectral sequences.** — Let $\mathbf{A}$ be an abelian category with enough projective objects. For any object $A$ and a cochain complex $C^* = (C^0 \to C^1 \to \cdots)$ of $\mathbf{A}$ one defines $\mathrm{Ext}^*_{\mathbf{A}}(A, C^*)$ to be homology of the total complex of the bicomplex $\mathrm{Hom}_{\mathbf{A}}(P_*, C^*)$, where $P_*$ is a projective resolution of $A$. There exist two spectral sequences, called respectively the first and the second hypercohomology spectral sequences. Both of them abut to the group $\mathrm{Ext}^*_{\mathbf{A}}(A, C^*)$. The first one has the form:

$$\mathbf{I}_2^{pq} = \mathrm{Ext}^q_{\mathbf{A}}(A, C^q) \Longrightarrow \mathrm{Ext}^{p+q}_{\mathbf{A}}(A, C^*)$$

while the second one has the form:

$$\mathbf{II}^{pq} = \mathrm{Ext}^p_{\mathbf{A}}(A, \mathrm{H}^q(C^*)) \Longrightarrow \mathrm{Ext}^{p+q}_{\mathbf{A}}(A, C^*).$$

We are going to use these spectral sequences in the case, when $\mathbf{A} = \mathcal{F}$ and $A = \mathrm{Id}$.

By Section 1.7 we know that the de Rham complex gives rise to the following cochain complex:

$$\Omega^*_n = (S^n \xrightarrow{d} S^{n-1} \otimes \mathrm{Id} \xrightarrow{d} \cdots \xrightarrow{d} \mathrm{Id} \otimes \Lambda^{n-1} \xrightarrow{d} \Lambda^n \longrightarrow 0 \longrightarrow 0 \longrightarrow \cdots).$$

Thanks to Lemma 1.2 we know that

(11) $$\mathrm{H}^*(\Omega^*_{p^h}) \cong \Omega^*_{p^{h-1}}.$$

**Proposition 3.5.** — *The Frobenius transformation* $\Phi : \mathrm{S}^{p^h} \to \mathrm{S}^{p^{h+1}}$ *yields an isomorphism*

$$\Phi_* : \mathrm{H}^i(\mathrm{S}^{p^{h-1}}) \cong \mathrm{H}^i(\mathrm{S}^{p^h}) \quad \text{if } 0 \leqslant i \leqslant 2p^{h-1} - 2.$$

*Proof.* — The first hypercohomology spectral sequence corresponding to the cochain complex $\Omega^*_{p^h}$ has the form $_\Omega\mathbf{I}_1^{st} = \mathrm{H}^t(\Lambda^s \otimes \mathrm{S}^{p^h-s})$. It follows from Lemma 2.13 and from (ii) of Lemma 3.4 that

$$_\Omega\mathbf{I}_1^{st} = \begin{cases} \mathrm{H}^t(\mathrm{S}^{p^h}) & \text{if } s = 0, \\ \mathrm{H}^{t-p^h+1}(\mathrm{S}^{p^h}) & \text{if } s = p^h, \\ 0 & \text{otherwise.} \end{cases}$$

The second hypercohomology spectral sequence corresponding to the bicomplex $\Omega^*_{p^h}$ has the form $_\Omega\mathbf{II}_2^{st} = \mathrm{H}^s(\mathrm{H}^t(\Omega^*_{p^h}))$. Thanks to the isomorphism (11), it has the following form:

$$_\Omega\mathbf{II}_2^{st} = \begin{cases} \mathrm{H}^s(\mathrm{S}^{p^{h-1}}) & \text{if } t = 0, \\ \mathrm{H}^{s-p^{h-1}+1}(\mathrm{S}^{p^{h-1}}) & \text{if } t = p^{h-1}, \\ 0 & \text{otherwise.} \end{cases}$$

Both spectral sequences have the same abutment $\mathrm{H}^*(\Omega^*_{p^h})$. The first spectral sequence gives the isomorphism: $\mathrm{H}^i(\mathrm{S}^{p^h}) \cong \mathrm{H}^*(\Omega^*_{p^h})$ for $i \leqslant 2p^h - 2$, while the second one gives $\mathrm{H}^i(\mathrm{S}^{p^{h-1}}) \cong \mathrm{H}^*(\Omega^*_{p^h})$ for $i \leqslant 2p^{h-1} - 2$ and we get the result. $\qquad \square$

**Corollary 3.6**. — *The group $\mathrm{H}^0(\mathrm{S}^{p^h})$ is one dimensional and the iterated Frobenius homomorphism is a generator. Moreover if $1 \leqslant i \leqslant 2p^h - 2$, then $\mathrm{H}^i(\mathrm{S}^{p^h}) = 0$.*

*Proof.* — It is a direct consequence of Lemma 3.3 and of Proposition 3.5. $\qquad \square$

**3.3. An auxiliary complex.** — In order to continue the computation of $\mathrm{H}^*(\mathrm{S}^*)$ we need an auxiliary complex $K_n^*$. It is defined as follows. First we set

$$K_n^i = \mathrm{Ker}(\kappa : \mathrm{S}^{n-i} \otimes \Lambda^i \longrightarrow \mathrm{S}^{n-i+1} \otimes \Lambda^{i-1}).$$

Thus $K_n^i = 0$ for $i \geqslant n$, $K_n^{n-1} = \Lambda^n$ and $K_n^0 = \mathrm{S}^n$. By exactness of the Koszul complex there is an exact sequence

$$0 \longrightarrow K_n^i \longrightarrow \mathrm{S}^{n-i} \otimes \Lambda^i \longrightarrow K_n^{i-1} \longrightarrow 0.$$

By Corollary 2.13 one gets

(12) $$\mathrm{H}^m(K_n^i) \cong \mathrm{H}^{m-i}(\mathrm{S}^n).$$

The Euler formula shows that the de Rham differential sends $K_m^i$ to $K_m^{i+1}$ provided $p \mid m$. Hence $K_{pn}$ is a subcomplex of $\Omega_{pn}$.

**Lemma 3.7**. — *The Cartier homomorphism yields an isomorphism:*

$$K_n^* \cong \mathrm{H}^*(K_{pn}^*, d).$$

*Proof.* — We will use the following well known fact from homological algebra: Let

$$X^* = (X^0 \longrightarrow X^1 \longrightarrow X^2 \longrightarrow \cdots)$$

be a cochain complex in an abelian category $\mathcal{A}$. If for each $k$ both $X^k$ and $\mathrm{H}^k(X^*)$ are injective objects in $\mathcal{A}$, then for any $A \in \mathcal{A}$ there is a natural isomorphism

$$\mathrm{H}^*(\mathrm{Hom}_{\mathcal{A}}(A, X^*)) \cong \mathrm{Hom}_{\mathcal{A}}(A, \mathrm{H}^*(X^*)).$$

One takes $\mathcal{A}$ to be the category of graded modules over the graded ring $D = \mathbb{F}_p[\varepsilon]$, where $\varepsilon^2 = 0$ and degree of $\varepsilon$ is $-1$. A graded module over $D$ is nothing else but a chain complex. It follows from the Euler formula that the de Rham differential yields the following cochain complex in $\mathcal{A}$

$$X^* = (X^0 \longrightarrow X^1 \longrightarrow X^2 \longrightarrow \cdots)$$

where $X_*^k = (\Omega_{pn}^{k-*}, (-1)^k \kappa)$. Then $X^k$ is an injective object in $\mathcal{A}$, because the Koszul complex is acyclic. The Cartier isomorphism shows that the same is true for $\mathrm{H}^k(X^*)$. Let us observe that for any object $(C_*, d)$ in $\mathcal{A}$ one has

$$\mathrm{Hom}_D(\mathbb{F}_p, C_*) \cong \mathrm{Ker}(d : C_0 \longrightarrow C_{-1}),$$

where $\mathbb{F}_p$ is concentrated in degree zero and the action of $\varepsilon$ on $\mathbb{F}_p$ is trivial. Therefore we have $\mathrm{Hom}_{\mathcal{A}}(\mathbb{F}_p, X^*) \cong K_{pn}^*$. Now we can write

$$\mathrm{H}^*(K_{pn}^*, d) \cong \mathrm{H}^*(\mathrm{Hom}_D(\mathbb{F}_p, X^*)) = \mathrm{Hom}_D(\mathbb{F}_p, \mathrm{H}^*(X^*)) \cong \mathrm{Hom}_D(\mathbb{F}_p, \Omega_n^*) = K_n^*. \qquad \square$$

**3.4. Proof of Theorem 3.1.** — Consider the first hypercohomology spectral sequence corresponding to the complex $K_{p^h}$. It has the following form: $_K\mathbf{I}_1^{st} = \mathrm{H}^t(K_{p^h}^s)$. By (12) we have

$$_K\mathbf{I}_1^{st} = \begin{cases} \mathrm{H}^{t-s}(\mathrm{S}^{p^h}) & \text{if } 0 \leqslant s \leqslant p^h - 1, \\ 0 & \text{otherwise.} \end{cases}$$

The second hypercohomology spectral sequence corresponding to the complex $K_{p^h}$ has the following form: $_K\mathbf{II}_2^{st} = \mathrm{H}^s(\mathrm{H}^t(K_{p^h}^*))$. By Lemma 3.7 we have

$$_K\mathbf{II}_2^{st} = \begin{cases} \mathrm{H}^{s-t}(\mathrm{S}^{p^{h-1}}) & \text{if } 0 \leqslant t \leqslant p^{h-1} - 1, \\ 0 & \text{otherwise.} \end{cases}$$

Both of them have the same abutment $\mathrm{H}^*(K_{pn})$. The inclusion of cochain complexes $K_{p^h}^* \hookrightarrow \Omega_{p^h}^*$ yields a morphism of spectral sequences

$$\gamma :_K \mathbf{II}_r^{st} \longrightarrow _\Omega\mathbf{II}_r^{st}.$$

Observe that $\gamma$ is the identity on the line $t = 0$ and $\gamma$ is zero on the terms $t > 0$, by a degree argument. Hence all differentials of the spectral sequence $_K\mathbf{II}_r^{st}$ which land on the line $t = 0$ are zero and hence $_K\mathbf{II}_2^{s0} = \mathrm{H}^*(\mathrm{S}^{p^{h-1}}) \to \mathrm{H}^*(K_{p^h})$ is a monomorphism.

Now we prove that for odd $k$ one has $\mathrm{H}^k(\mathrm{S}^{p^h}) = 0$. Indeed, fix a natural number $N$ and consider the following properties

$$(\mathcal{P}_N(h)) \qquad\qquad \mathrm{H}^k(\mathrm{S}^{p^h}) = 0 \quad \text{for odd } k \leqslant N,$$

$$(\mathcal{Q}_N(h)) \qquad\qquad \mathrm{H}^k(K_{p^h}^*) = 0 \quad \text{for odd } k \leqslant N.$$

Observe that for a fixed integer $N$ the property $(\mathcal{P}_N(h))$ holds for $h \gg 0$ (thanks to Corollary 3.6) and $(\mathcal{P}_N(h)) \Rightarrow (\mathcal{Q}_N(h))$ (because, if $k = s + t \leqslant N$ is odd, then $s - t$ is also odd and therefore in the total degree $k$ the terms of $_K\mathbf{I}$ are zero). On the other hand, we know that $\mathrm{H}^*(\mathrm{S}^{p^{h-1}}) \to \mathrm{H}^*(K_{p^h})$ is a monomorphism, which yields $(\mathcal{Q}_N(h)) \Rightarrow (\mathcal{P}_N(h-1))$. Thus $(\mathcal{P}_N(h))$ is true for any $h \geqslant 0$. Since $N$ was arbitrary, we see that $\mathrm{H}^k(\mathrm{S}^{p^h}) = 0$ for odd $k$.

To finish the computation let us recall that all terms of odd total degrees of the spectral sequences $_K\mathbf{I}$ and $_K\mathbf{II}$ are zero. Thus they die at $_K\mathbf{I}_1$ and $_K\mathbf{II}_2$. We denote by $P_h(X)$ the Poincaré series for $\mathrm{H}^*(\mathrm{S}^{p^h})$. Since $_K\mathbf{I}$ and $_K\mathbf{II}$ have the same abutment, one can compare their Poincaré series and we get:

$$\sum_{\alpha=0}^{p^h-1} X^{2\alpha} P_h(X) = \sum_{\alpha=0}^{p^{h-1}-1} X^{2\alpha} P_{h-1}(X).$$

Therefore, for any $h$ one has:

$$(13) \qquad P_0(X) = \sum_{\alpha=0}^{p^h-1} X^{2\alpha} P_h(X).$$

It follows from Corollary 3.6, that the $< N$ degree part of $P_h(X)$ is 1 if $h \gg 0$. This fact, together with equation (13), shows that $P_0(X) = \sum_{\alpha \geqslant 0} X^{2\alpha}$. Applying (13) again finishes the proof. $\qquad\qquad\qquad\qquad\qquad\qquad\qquad\qquad\qquad\qquad\qquad\quad$ □

## 4. Polynomial functors à la Friedlander and Suslin

**4.1. Strict polynomial functors.** — In this section $\mathbb{K}$ is a field. In this section we introduce *strict polynomial functors* [**9**]. These objects first appear in the unpublished work of Bousfield [**2**]. In order to define strict polynomial functors we need an additional category $\Gamma^d \mathcal{V}^f$.

For any vector space $V$, there is a map $\gamma^d : V \to \Gamma^d(V)$, given by: $\gamma^d(x) := x^{\otimes d} \in (V^{\otimes d})^{\Sigma_d}$. For any vector spaces $V$ and $W$ there exists a natural linear map $\mathrm{S}^d(V \otimes W) \to \mathrm{S}^d(V) \otimes \mathrm{S}^d(W)$ given by $(x_1 \otimes y_1) \cdots (x_d \otimes y_d) \mapsto (x_1 \cdots x_d) \otimes (y_1 \cdots y_d)$. The duality between symmetric and divided powers yields a natural map

$$\mu : \Gamma^d V \otimes \Gamma^d W \longrightarrow \Gamma^d(V \otimes W).$$

One easily checks that $\mu(\gamma^d(x) \otimes \gamma^d(y)) = \gamma^d(x \otimes y)$. Building on the transformation $\mu$ one can define the category $\Gamma^d \mathcal{V}^f$ as follows. Objects of $\Gamma^d \mathcal{V}^f$ are the same as of $\mathcal{V}^f$, while morphisms in $\Gamma^d \mathcal{V}^f$ are

$$\mathrm{Hom}_{\Gamma^d \mathcal{V}^f}(V, W) := \Gamma^d(\mathrm{Hom}(V, W)).$$

The identity arrow in $\Gamma^d \mathcal{V}^f$ corresponding to an object $V$ is $\gamma^d(1_V)$. Composition in $\Gamma^d \mathcal{V}^f$ is induced by the composition in $\mathcal{V}^f$ and by the transformation $\mu$

$$\Gamma^d(\mathrm{Hom}(V, W)) \otimes \Gamma^d(\mathrm{Hom}(U, V)) \xrightarrow{\mu} \Gamma^d(\mathrm{Hom}(V, W) \otimes \mathrm{Hom}(U, V))$$

$$\circ \searrow \qquad\qquad\qquad \downarrow$$

$$\Gamma^d \, \mathrm{Hom}(U, W).$$

The category $\Gamma^d \mathcal{V}$ is a $\mathbb{K}$-linear category, that is a category whose set of morphisms between two objects has a $\mathbb{K}$-vector space structure and the composition is bilinear. A *homogeneous strict polynomial functor of degree $d$* is a $\mathbb{K}$-linear functor $T : \Gamma^d \mathcal{V}^f \to \mathcal{V}^f$. In other words $T$ is a functor and the structural maps

$$\Gamma^d(\mathrm{Hom}(V, W)) \longrightarrow \mathrm{Hom}(T(V), T(W))$$

are $\mathbb{K}$-linear. The definition of strict polynomial functors is equivalent to one given in [**8**] thanks to [**8**, Remark 3.7].

We let $\mathcal{P}_d$ be the category of strict homogeneous polynomial functors of degree $d$. We put $\mathcal{P} = \bigoplus_d \mathcal{P}_d$, which is by definition a full subcategory of $\prod_d \mathcal{P}_d$ of objects which are zero except for a finite number of components. An object $T$ of $\mathcal{P}$ is called a *strict polynomial functor*. By abuse of notation we write $T = \bigoplus_{d \geqslant 0} T_d$, where $T_d \in \mathcal{P}_d$. One puts

$$\operatorname{Deg}(T) := \sup_d \{d \mid T_d \neq 0\}$$

For each $m \geqslant 0$ we let $\Gamma^{d,m} \in \mathcal{P}_d$ be the functor given by

$$\Gamma^{d,m}(V) = \Gamma^d(\operatorname{Hom}(\mathbb{K}^m, V)).$$

By the Yoneda lemma there is an isomorphism $\operatorname{Hom}_{\mathcal{P}}(\Gamma^{d,m}, F) \cong F(\mathbb{K}^m)$ for any $F \in \mathcal{P}_d$. So the $\Gamma^{d,m}$, $m \geqslant 0$ are small projective generators in the abelian category $\mathcal{P}_d$. Using cross-effects it is not too difficult to show that $\Gamma^{d,n}$ is a projective generator provided $n \geqslant d$. Thus $\mathcal{P}_d$ is equivalent to the category of finite dimensional modules over $\operatorname{End}_{\mathcal{P}}(\Gamma^{d,n}) = \Gamma^d(\operatorname{End}(\mathbb{K}^n))$, if $n \geqslant d$. The algebra $\Gamma^d(\operatorname{End}(\mathbb{K}^n))$ is known as the *Schur algebra* $S(d,n)$ [**9**].

**Remark 4.1**. — If $\mathbb{K}$ is an infinite field, homogeneous strict polynomial functors of degree $d$ can be identified with functors $T : \mathcal{V}^f \to \mathcal{V}^f$ such that the structural maps $\operatorname{Hom}(V, W) \to \operatorname{Hom}(TV, TW)$ are morphisms of algebraic varieties given by homogeneous polynomials of degree $d$. This is no longer true if $\mathbb{K}$ is a finite field, because over such fields there is an essential difference between polynomials and maps given by polynomials (see the introduction to this volume).

Clearly $f \mapsto \gamma^d(f)$ defines a (nonlinear) functor $\gamma^d \colon \mathcal{V} \to \Gamma^d \mathcal{V}$. Precomposition with $\gamma^d$ yields a functor $\mathcal{P}_d \to \mathcal{F}$, which is a full embedding provided the field $K$ contains at least $d$ elements (see [**9**]). By abuse of notation we denote the image of $T \in \mathcal{P}_d$ under this functor by the same letter $T$. Then we have $\deg(T) \leqslant \operatorname{Deg}(T)$, in other words the functor $\gamma^d$ yields a functor $\mathcal{P}_d \to \mathcal{F}_d$.

**Remark 4.2**. — Let us observe that objects of $\mathcal{P}_0$ and $\mathcal{F}_0$ are just constant functors. For $d > 0$ the categories $\mathcal{P}_d$ and $\mathcal{F}_d$ are quite different. For the category $\mathcal{F}_d$ one has a decomposition $\mathcal{F}_d \cong \mathcal{F}_0 \times \overline{\mathcal{F}_d}$, where $\overline{\mathcal{F}_d}$ consists with functors $F$ such that $F(0) = 0$. If $d = 1$, then $\mathcal{P}_1$ is equivalent to the category $\mathcal{V}^f$, while $\overline{\mathcal{F}_1}$ is equivalent to the category of $\mathbb{K}$-$\mathbb{K}$-bimodules. Let $\mathfrak{S}_d$-Rep be the category of finite dimensional representations of $\mathfrak{S}_d$. There is a functor $\mathcal{P}_d \to \mathfrak{S}_d$-Rep, which assigns the value of $d$-th cross-effect (see [**7**]) of $T$ on $(\mathbb{K}, \ldots, \mathbb{K})$ to an object $T \in \mathcal{P}_d$. If $\mathbb{K}$ is a field of characteristic zero, then this is an equivalence of categories (see [**11**]). The inverse functor assigns to a representation $M$ the functor which is given by $X \mapsto M \otimes_{\mathfrak{S}_d} X^{\otimes d}$. Similarly, if $\mathbb{K} = \mathbb{Q}$ is the field of rational numbers, then one has an equivalence of categories $\mathcal{F}_d \cong \prod_{n=1}^{d} (\mathbb{Q}[\mathfrak{S}_n]\text{-Mod})$.

The natural transformation $\Gamma^{d+l} \to \Gamma^d \otimes \Gamma^l$ coming from the Hopf algebra structure on the divided power algebra can be used to define the tensor product of strict homogeneous polynomial functors. It yields a biexact functor $-\otimes- : \mathcal{P}_d \times \mathcal{P}_l \to \mathcal{P}_{k+l}$, which corresponds to the usual tensor functor in $\mathcal{F}$.

Multiplication in the symmetric algebra yields a natural transformation $S^m \circ S^n \to S^{mn}$. By duality we get the transformation $\Gamma^{mn} \to \Gamma^m \circ \Gamma^n$, which can be used to define the composition of strict homogeneous polynomial functors $- \circ - : \mathcal{P}_n \times \mathcal{P}_m \to \mathcal{P}_{mn}$. One observes also that the dual of a strict homogeneous polynomial functor is a strict homogeneous polynomial functor. Since $\Gamma^d$ carries an obvious structure of a strict homogeneous polynomial functor of degree $d$, we see that $T^d, S^d, \Lambda^d$ are in $\mathcal{P}_d$. The functors $S^{d,m} := (\Gamma^{d,m})^\natural$, $m \geqslant 0$, are injective objects in $\mathcal{P}_d$, because they are dual objects of projective ones.

If $\mathbb{K}$ is a field of characteristic $p > 0$, one can introduce the Frobenius twist as follows. For any $F \in \mathcal{P}_d$ we put $F^{(1)} = F \circ \mathrm{Id}^{(1)}$ and then by induction one defines $F^{(r+1)} = (F^{(r)})^{(1)}, r \geqslant 0$. By definition $F^{(r)} \in \mathcal{P}_{dp^r}$ if $F \in \mathcal{P}_d$. In particular $\mathrm{Id}^{(1)}$ belongs to $\mathcal{P}_p$ and is not isomorphic to $\mathrm{Id} \in \mathcal{P}_1$ even for finite prime fields $\mathbb{F}_p$, though for such a field both $\mathrm{Id}^{(1)}$ and $\mathrm{Id}$ give rise to the same functor in $\mathcal{F}$.

Let $F, T \in \mathcal{P}_d$. Then the Frobenius twist gives a homomorphism

$$\mathrm{Ext}^i_{\mathcal{P}}(F, T) \longrightarrow \mathrm{Ext}^i_{\mathcal{P}}(F^{(1)}, T^{(1)}).$$

We let $\mathrm{Ext}^i_{\mathrm{Gen}}(F, T)$ be the corresponding colimit:

$$\mathrm{Ext}^i_{\mathrm{Gen}}(F, T) := \mathrm{colim}_m \mathrm{Ext}^i_{\mathcal{P}}(F^{(m)}, T^{(m)}).$$

From now on we assume that $\mathbb{K}$ is a finite field. Since the Frobenius morphism $\mathbb{K} \to \mathbb{K}$ is an isomorphism, the Frobenius twist in $\mathcal{F}$ is an equivalence of categories. We see that the canonical map $\mathrm{Ext}^*_{\mathcal{P}}(F, T) \to \mathrm{Ext}^*_{\mathcal{F}}(F, T)$ gives rise to the homomorphism

$$\mathrm{Ext}^*_{\mathrm{Gen}}(F, T) \longrightarrow \mathrm{Ext}^*_{\mathcal{F}}(F, T).$$

In general $\mathrm{Ext}^*_{\mathrm{Gen}}(F, T) \to \mathrm{Ext}^*_{\mathcal{F}}(F, T)$ is not an isomorphism even when $* = 0$. The following rather surprising result of Franjou, Friedlander, Scorichenko and Suslin gives a condition when it is an isomorphism.

**Theorem 4.3** ([5]). — *Let $T, F \in \mathcal{P}_d$ be strict polynomial functors, then*

$$\mathrm{Ext}^*_{\mathcal{P}}(F, T) \longrightarrow \mathrm{Ext}^*_{\mathcal{F}}(F, T)$$

*is a monomorphism. Moreover if the field $K$ contains at least $d$ elements then*

$$\mathrm{Ext}^*_{\mathrm{Gen}}(F, T) \longrightarrow \mathrm{Ext}^*_{\mathcal{F}}(F, T)$$

*is an isomorphism.*

The proof of this theorem is rather complicated and very long and we refer the reader to the original paper. Instead in the next section we give the computation of $\mathrm{Ext}^*_{\mathcal{P}}(\mathrm{Id}^{(r)}, S^{n(j)})$, which is a starting point for Theorem 4.3.

**4.2. Computations in $\mathcal{P}$.** — In this section we will compute $\mathrm{Ext}_{\mathcal{P}}^*(\mathrm{Id}^{(r)}, \mathrm{S}^{n(j)})$. Since $\mathrm{Id}^{(r)} \in \mathcal{P}_{p^r}$ and $\mathrm{S}^{n(j)} \in \mathcal{P}_{np^j}$ we see that the groups in question are zero by trivial reasons provided $n \neq p^{r-j}$. Thus we will assume that $n = p^{r-j}$. The main technical tool is the following lemma which is completely similar to Corollary 2.13.

**Lemma 4.4.** — *Let $A, F, T \in \mathcal{P}$. If $A$ is additive considered as an object of the category $\mathcal{F}$ and $T(0) = 0 = F(0)$, then*

$$\mathrm{Ext}_{\mathcal{P}}^*(A, T \otimes F) = 0.$$

The lemma permits us to use the same strategy as the one given in Section 3. We are going to prove the following

**Theorem 4.5 ([9]).** — *Let $n = p^{r-j}$. Then we have*

$$\mathrm{Ext}_{\mathcal{P}_d}^i(\mathrm{Id}^{(r)}, \mathrm{S}^{n(j)}) = \begin{cases} \mathbb{K}, & \text{if } 2n \mid i \text{ and } i < 2p^r \\ 0 & \text{otherwise.} \end{cases}$$

*Proof.* — By induction on $j$. If $j = 0$ the result is clear, because $\mathrm{S}^{n(j)} = \mathrm{S}^n$ is injective and $\mathrm{Hom}_{\mathcal{P}}(I^{(r)}, \mathrm{S}^{p^r})$ is spanned by the iterated Frobenius $\Phi^r$. We will assume that the statement is true for $j$ and we will prove it for $j + 1$. Then we obtain (compare with (ii) of Proposition 3.4):

$$\mathrm{Ext}_{\mathcal{P}_d}^i(\mathrm{Id}^{(r)}, \Lambda^{n(j)}) = \begin{cases} \mathbb{K}, & \text{if } i \equiv n - 1 \,(\mathrm{mod}\,2\mathrm{n}) \text{ and } i < 2p^r \\ 0 & \text{otherwise.} \end{cases}$$

Indeed, one needs only to use the $j$-th twist of the Koszul exact sequence:

$$0 \longrightarrow \Lambda^{n(j)} \longrightarrow \cdots \longrightarrow \Lambda^{(n-i)(j)} \otimes \mathrm{S}^{i(j)} \longrightarrow \cdots \longrightarrow \mathrm{S}^{n(j)} \longrightarrow 0$$

and Lemma 4.4. Now we consider the hypercohomology spectral sequence for the $j$-th twist of the de Rham complex:

$$\Omega_n^{*(j)} = (\mathrm{S}^{n(j)} \longrightarrow \cdots \longrightarrow \mathrm{S}^{(n-i)(j)} \otimes \Lambda^{i(j)} \longrightarrow \cdots \longrightarrow \Lambda^{n(j)} \longrightarrow 0 \longrightarrow \cdots)$$

The first hypercohomology spectral sequence has the form

$$_{\Omega_n^{*(j)}}\mathbf{I}_1^{st} = \mathrm{Ext}_{\mathcal{P}}^t(\mathrm{Id}^{(r)}, \mathrm{S}^{n-s(j)} \otimes \Lambda^{s(j)}) \Longrightarrow \mathrm{Ext}_{\mathcal{P}}^{s+t}(\mathrm{Id}^{(r)}, \Omega_n^{*(j)}).$$

By Lemma 4.4 we have only two nontrivial columns corresponding to $s = 0$ and $s = n$:

$$_{\Omega_n^{*(j)}}\mathbf{I}_1^{0t} = \mathrm{Ext}_{\mathcal{P}}^t(\mathrm{Id}^{(r)}, \mathrm{S}^{n(j)})$$

$$_{\Omega_n^{*(j)}}\mathbf{I}_1^{nt} = \mathrm{Ext}_{\mathcal{P}}^t(\mathrm{Id}^{(r)}, \Lambda^{n(j)}) \cong \mathrm{Ext}_{\mathcal{P}}^{t-n+1}(\mathrm{Id}^{(r)}, \mathrm{S}^{n(j)})$$

By our assumptions we know all nontrivial terms: they are one dimensional vector spaces concentrated in bidegrees $(0, 2n), \ldots, (0, 2d - 2n), (n, n - 1), \ldots (n, 2d - n - 1)$. Therefore there are no differentials by degree reasons. Thus

$$\mathrm{Ext}_{\mathcal{P}}^i(\mathrm{Id}^{(r)}, \Omega_n^{*(j)}) = \begin{cases} \mathbb{K}, & \text{if } i \equiv 0, -1 \,(\mathrm{mod}\,2\mathrm{n}), \text{ and } i < 2p^r, \\ 0 & \text{otherwise.} \end{cases}$$

Now we consider the second hypercohomology spectral sequence associated to the complex $\Omega^{*(j)}$:

$$_{\Omega_n^{*(j)}}\mathbf{II}_2^{st} = \operatorname{Ext}_{\mathcal{P}}^s(\operatorname{Id}^{(r)}, \operatorname{H}^t(\Omega_n^{*(j)})) \Longrightarrow \operatorname{Ext}_{\mathcal{P}}^{s+t}(\operatorname{Id}^{(r)}, \Omega_n^{*(j)}).$$

By the result of Cartier we know that

$$\operatorname{H}^*(\Omega_n^{*(j)}) \cong \Omega_m^{*(j+1)},$$

where $m = p^{r-j-1}$. It follows again from Corollary 2.13 that the spectral sequence has only two nontrivial rows $t = 0$ and $t = m$:

$$_{\Omega_n^{*(j)}}\mathbf{II}_2^{s0} = \operatorname{Ext}_{\mathcal{P}}^s(\operatorname{Id}^{(r)}, \operatorname{S}^{m(j+1)})$$

$$_{\Omega_n^{*(j)}}\mathbf{II}_2^{sm} = \operatorname{Ext}_{\mathcal{P}}^s(\operatorname{Id}^{(r)}, \Lambda^{m(j+1)}) \cong \operatorname{Ext}_{\mathcal{P}}^{s-m+1}(\operatorname{Id}^{(r)}, \operatorname{S}^{m(j+1)}).$$

Thus it gives rise to a long exact sequences:

$$\cdots \longrightarrow \operatorname{Ext}_{\mathcal{P}}^i(\operatorname{Id}^{(r)}, \operatorname{S}^{m(j+1)}) \longrightarrow \operatorname{Ext}_{\mathcal{P}}^i(\operatorname{Id}^{(r)}, \Omega_n^{*(j)}) \longrightarrow \operatorname{Ext}_{\mathcal{P}}^{i-2m+1}(\operatorname{Id}^{(r)}, \operatorname{S}^{m(j+1)})$$

$$\longrightarrow \operatorname{Ext}_{\mathcal{P}}^{i+1}(\operatorname{Id}^{(r)}, \operatorname{S}^{m(j+1)}) \longrightarrow \operatorname{Ext}_{\mathcal{P}}^{i+1}(\operatorname{Id}^{(r)}, \Omega_n^{*(j)}) \longrightarrow \cdots$$

Since the groups $\operatorname{Ext}_{\mathcal{P}}^*(\operatorname{Id}^{(r)}, \Omega_n^{*(j)})$ are known, it suffices to prove that for odd $i$ one has

$$\operatorname{Ext}_{\mathcal{P}}^i(\operatorname{Id}^{(r)}, \operatorname{S}^{m(j+1)}) = 0$$

To this end we consider the complex $K_n^*$ which is the obvious modification in the world of strong polynomial functors of the complex considered in Section 3.3. We have $\operatorname{H}^*(K_n^*) \cong K_m^{*(1)}$, where as above $m = p^{r-j-1}$ and $n = p^{r-j}$. The first hypercohomology spectral sequence corresponding to the complex $K_n^{*(j)}$ has the following form

$$_{K_n^{*(j)}}\mathbf{I}_1^{st} = \operatorname{Ext}_{\mathcal{P}}^t(\operatorname{Id}^{(r)}, K_n^{s(j)}) = \begin{cases} \operatorname{Ext}_{\mathcal{P}}^{t-s}(\operatorname{Id}^{(r)}, \operatorname{S}^{n(j)}) & \text{if } 0 \leqslant s \leqslant n-1, \\ 0 & \text{otherwise.} \end{cases}$$

By induction assumption the groups $\operatorname{Ext}_{\mathcal{P}}^{t-s}(\operatorname{Id}^{(r)}, \operatorname{S}^{n(j)})$ are known. We can deduce that $\operatorname{Ext}_{\mathcal{P}}^i(\operatorname{Id}^{(r)}, K_n^{*(j)}) = 0$ provided $i$ is odd.

The second hypercohomology spectral sequence corresponding to the complex $K_n^{*(j)}$ has the following form

$$_{K_n^{*(j)}}\mathbf{II}_2^{st} = \operatorname{Ext}_{\mathcal{P}}^s(\operatorname{Id}^{(r)}, \operatorname{H}^t(K_n^{*(j)})) = \begin{cases} \operatorname{Ext}_{\mathcal{P}}^{s-t}(\operatorname{Id}^{(r)}, \operatorname{S}^{m(j+1)}) & \text{if } 0 \leqslant t \leqslant m-1, \\ 0 & \text{otherwise.} \end{cases}$$

The inclusion of cochain complexes $K_n^{*(j)} \hookrightarrow \Omega_n^{*(j)}$ yields a morphism of spectral sequences

$$\gamma : _{K_n^{*(j)}}\mathbf{II}_r^{st} \longrightarrow _{\Omega_n^{*(j)}}\mathbf{II}_r^{st}$$

Observe that $\gamma$ is the identity on the line $t = 0$ and $\gamma$ is zero on the terms $t > 0$, by a degree argument. Hence all differentials of the spectral sequence $_{K_n^{*(j)}}\mathbf{II}_r^{st}$ which land on the line $t = 0$ are zero. Thus $_{K_n^{*(j)}}\mathbf{II}_2^{s0} = \operatorname{Ext}_{\mathcal{P}}^*(\operatorname{Id}^{(r)}, \operatorname{S}^{m(j+1)}) \to \operatorname{Ext}_{\mathcal{P}}^i(\operatorname{Id}^{(r)}, K_n^{*(j)})$

is a monomorphism. It follows that $\operatorname{Ext}^i_{\mathcal{P}}(\operatorname{Id}^{(r)}, S^{m(j+1)}) = 0$ if $i$ is odd and hence the result.                                                                                    $\square$

# References

[1] R. BIERI – *Homological dimension of discrete groups*, Queen Mary College Mathematics Notes, Mathematics Department, Queen Mary College, London, 1976.

[2] A.K. BOUSFIELD – *Homogeneous functors and their derived functors*, Brandeis University, 1967.

[3] L. BREEN – Extensions du groupe additif, *Publ. Math. Inst. Hautes Études Sci.* **48** (1978), p. 39–125.

[4] P. CARTIER – Une nouvelle opération sur les formes différentielles, *C. R. Acad. Sci. Paris Sér. I Math.* **244** (1957), p. 426–428.

[5] V. FRANJOU, E.M. FRIEDLANDER, A. SCORICHENKO & A. SUSLIN – General linear and functor cohomology over finite fields, *Ann. of Math. (2)* **150** (1999), p. 663–728.

[6] V. FRANJOU, J. LANNES & L. SCHWARTZ – Autour de la cohomologie de MacLane des corps finis, *Invent. Math.* **115** (1994), no. 3, p. 513–538.

[7] V. FRANJOU & T. PIRASHVILI – Stable $K$-theory is bifunctor homology (after A. Scorichenko), in this volume.

[8] E.M. FRIEDLANDER – Lectures on the cohomology of finite group schemes, in this volume.

[9] E.M. FRIEDLANDER & A. SUSLIN – Cohomology of finite group schemes over a field, *Invent. Math.* **127** (1997), p. 209–270.

[10] J.-L. LODAY – *Cyclic homology*, 2nd ed., Grundlehren der Mathematischen Wissenschaften, vol. 301, Springer-Verlag, Berlin, 1998.

[11] I.G. MACDONALD – *Symmetric functions and Hall polynomials*, 2nd ed., Oxford Mathematical Monographs, Oxford University Press, 1995.

[12] T. PIRASHVILI – Higher additivizations, *Trudy Tbiliss. Mat. Inst. Razmadze Akad. Nauk Gruzin. SSR* **91** (1988), p. 44–54, Russian.

[13] N. ROBY – Lois polynômes et lois formelles en théorie des modules, *Ann. Sci. École Norm. Sup.* **80** (1963), p. 213–348.

[14] L. SCHWARTZ – Algèbre de Steenrod, modules instables et foncteurs polynomiaux, in this volume.

*Panoramas & Synthèses*
**16**, 2003, p. 27–53

# LECTURES ON THE COHOMOLOGY
# OF FINITE GROUP SCHEMES

*by*

Eric M. Friedlander

———————————

**Abstract**. — We provide an introduction to the cohomology of finite group schemes, a class of objects which includes finite groups and $p$-restricted Lie algebras. Various qualitative results, known earlier for finite groups by work of Quillen and others, are extended to this general context. Various computational techniques which arise from classical homological algebra are recalled. We then proceed to discuss the essential role of strict polynomial functors in the proof of the fundamental theorem which asserts that the cohomology of a finite group scheme is finitely generated.

***Résumé* (Cohomologie des schémas en groupes finis)**. — Ce texte est une introduction à la cohomologie des schémas en groupes finis. Cette classe d'objets contient les groupes finis et les algèbres de Lie restreintes. Plusieurs résultats qualitatifs, établis pour les groupes finis par Quillen et d'autres, leurs sont généralisés. On rappelle les méthodes de calcul de l'algèbre homologique, puis on explique l'intervention déterminante des foncteurs polynomiaux stricts dans la démonstration qui établit que la cohomologie d'un schéma en groupes fini est de type fini.

## 0. Introduction

The goal of these lectures (which were presented in a preliminary form at the Nantes meeting) is to provide an introduction to some of the techniques and computations of cohomology of finite group schemes over a field $k$ of characteristic $p > 0$ which have been developed since the publication of J. Jantzen's book [**13**] and to explain the important role played by the cohomology of (strict polynomial) functors. The focal point of these lectures is a theorem of E. Friedlander and A. Suslin asserting that the cohomology of finite group schemes is finitely generated (see Theorem 4.7 below). The somewhat innovative proof of this theorem has led to numerous further results; in these lectures we have restricted attention to those results bearing on the qualitative description of the cohomology algebra of a finite group scheme.

**2000** *Mathematics Subject Classification*. — 14L15, 20G10.
***Key words and phrases***. — Group scheme, Hopf algebra, algebraic group, restricted Lie algebra, general linear group, rational representation, Schur algebra, weights, support variety, finite generation.

The reader can obtain a quick guide to these edited lectures by glancing at the table of contents. In the first lecture, we introduce the concepts and terminology which underline our subject. In particular, we recall the definition of the Frobenius kernels of an algebraic group and the Frobenius twists of a module. The second lecture summarizes some of the techniques which one can find for example in [**13**] which are used to compute cohomology. The relationship of this subject with the theme of the Nantes meeting, cohomology in categories of functors, is explained in the third lecture. Strict polynomial functors are introduced and their relationship with polynomial representations is explained. The fourth lecture is dedicated to an outline of the proof of finite generation of the cohomology of finite group schemes. Here, computations of cohomology in the category of strict polynomial functors plays a central role in the construction of certain universal classes; these computations follow closely the computations of V. Franjou, J. Lannes, and L. Schwartz [**8**] of ordinary functor cohomology. Finally, in Lecture 5 we describe how the techniques introduced to prove finite generation lead to a qualitative description of the cohomology algebra $H^*(G, k)$ of a finite group scheme. This follows work of D. Quillen [**15**] who determined the maximal ideal spectrum of the cohomology of a finite group.

## 1. Affine group schemes

Let $k$ be a field of characteristic $p > 0$, fixed throughout this paper. We begin our discussion by defining an affine group scheme (implicitly assumed to be over $k$) and considering a few interesting examples.

***Definition 1.1***. — An affine group scheme is a representable functor
$$G : (\text{fin.gen.comm.}k\text{-}alg) \longrightarrow (grps)$$
We denote by $k[G]$ the representing finitely generated commutative $k$-algebra (the *coordinate algebra*) of $G$. To give such a representable functor is equivalent to giving a finitely generated commutative Hopf algebra (over $k$).

***Example 1.2***. — $G = \mathbb{G}_a$, the additive group. This is the functor which takes a commutative $k$-algebra $A$ to the underlying abelian group (which we might denote $A^+$). The coordinate algebra of $\mathbb{G}_a$ is $k[\mathbb{G}_a] = k[t]$, with coproduct $\Delta(t) = t \otimes 1 + 1 \otimes t$.

***Example 1.3***. — $G = \text{GL}_n$, the general linear group, sends a commutative $k$-algebra $A$ to the group of $n \times n$ invertible matrices $\{a_{i,j}\}$ with coefficients in $A$. The coordinate algebra of $\text{GL}_n$ is given by
$$k[\text{GL}_n] = k[x_{i,j}, t]_{1 \leqslant i,j \leqslant n} / \det(x_{i,j})t - 1$$
with coproduct
$$\Delta(x_{i,j}) = \Sigma x_{i,k} \otimes x_{k,j}.$$

***Example 1.4***. — Let $\pi$ be a (discrete) group. We view $\pi$ as an affine group scheme by letting $\pi$ also denote "the constant functor with value $\pi$." In other words, this functor sends a commutative $k$-algebra $A$ to the group $\pi^{|\pi_0(A)|}$, where $\pi_0(A)$ is the set of indecomposable non-trivial idempotents in $A$ and $|\pi_0(A)|$ denotes the cardinality of $\pi_0(A)$.

***Example 1.5***. — For any positive integer $r$, we consider the "$r$-th Frobenius kernel" of $\mathrm{GL}_n$ which is denoted $\mathrm{GL}_{n(r)}$. This is the functor which sends a commutative $k$-algebra $A$ to the group of $n \times n$ invertible matrices $(a_{i,j})$ with coefficients in $A$ which satisfy the property that $a_{i,j}^{p^r} = \delta_{i,j}$ (*i.e.*, equal to 1 if $i = j$ and 0 otherwise). The coordinate algebra $k[\mathrm{GL}_{n(r)}]$ is the quotient of $k[\mathrm{GL}_n]$ by the (Hopf) ideal generated by $x_{i,j}^{p^r} - \delta_{i,j}$. More explicitly, we can write $k[\mathrm{GL}_{n(r)}] = k[x_{i,j}]/(x_{i,j}^{p^r} - \delta_{i,j})$.

Similarly, the $r$-th Frobenius kernel of $\mathbb{G}_a$ sends $A$ to the group of elements of $A$ whose $p^r$-th power is 0. The coordinate algebra of $\mathbb{G}_{a(r)}$ is given by $k[\mathbb{G}_{a(r)}] = k[t]/t^{p^r}$, whereas the dual algebra is given by $k\mathbb{G}_{a(r)} = k[X_1, \ldots, X_r]/(X_i^p)$ where one can view the dual generator $X_i$ as the operator $\dfrac{1}{p^{i-1}!} \dfrac{d^{p^{i-1}}}{dt^{p^{i-1}}}$ on $k[t]$.

***Example 1.6***. — Let $g$ be a finite dimensional $p$-restricted Lie algebra of $k$ and let $V(g)$ denote its restricted enveloping algebra, the quotient of the universal enveloping algebra $U(g)$ of $g$ by the ideal generated by $\{X^p - X^{[p]}, X \in g\}$ (where $(-)^{[p]} : g \to g$ is the $p$-th power operation of $g$), Then the $k$-linear dual of $V(g)$, which we denote by $V(g)^{\#}$, is a finite dimensional commutative Hopf algebra over $k$ and thus corresponds to an affine group scheme over $k$.

***Remark 1.7***. — An affine group scheme $G$ is said to be *finite* if $k[G]$ is finite dimensional. For example, if $G$ corresponds to a finite group $\pi$ as in Example 1.4 or if $G$ is a group scheme as in Example 1.5 or $G$ is associated to a finite dimensional $p$-restricted Lie algebra as in Example 1.6, then $G$ is a finite group scheme. The linear dual is called the *group algebra* of $G$, denoted $kG$, consistent with the usual terminology of the group algebra of a discrete group $\pi$. In Example 1.6, the group algebra $kG$ of the group scheme $G$ associated to the $p$-restricted Lie algebra $g$ is $V(g)$, the restricted enveloping algebra of $g$.

One usually refers to an affine group scheme $G$ whose coordinate algebra is integral (*i.e.*, reduced and irreducible) as an (affine) algebraic group. For example, both $\mathbb{G}_a$ of Example 1.2 and $\mathrm{GL}_n$ of Example 1.3 are algebraic groups.

***Remark 1.8***. — A finite group scheme $G$ is said to be *infinitesimal* if the coordinate algebra $k[G]$ is local. An infinitesimal group $G$ is said to be of height $\leqslant r$ if $G$ admits a closed embedding $G \hookrightarrow \mathrm{GL}_{n(r)}$ (*i.e.*, if $a^{p^r} = 0$ for every element $a$ in the augmentation ideal of $k[G]$). For any infinitesimal group scheme $G$ of height 1 we have an isomorphism of algebras:

$$kG \simeq V(LieG).$$

Conversely, if $g$ is a finite dimensional $p$-restricted Lie algebra, then $V(g)^{\#}$ is the coordinate algebra of an infinitesimal group scheme $G$ of height 1. This establishes an equivalence of categories between finite dimensional $p$-restricted Lie algebras and infinitesimal group schemes of height 1.

We next introduce the concept of a $G$-module for an affine group scheme (sometimes called a rational $G$-module).

**Definition 1.9**. — Let $G$ be an affine group scheme over $k$. Then a $G$-module $M$ is a $k$-vector space provided with an $A$-linear group action

$$(1.10) \qquad\qquad G(A) \times (M \otimes A) \longrightarrow M \otimes A$$

for all finitely generated commutative $k$-algebras $A$, functorial with respect to $A$. (Here, and below, the tensor product is over $k$.)

Equivalently, such a $G$-module $M$ is a $k$-vector space provided with the structure of a comodule for $k[G]$; namely, a $k$-linear map

$$(1.11) \qquad\qquad \Delta_M : M \longrightarrow M \otimes k[G].$$

To verify this equivalence, observe that the pairing (1.10) in the special case $A = k[G]$ is written

$$\mathrm{Hom}_{k\text{-alg}}(k[G], k[G]) \times (M \otimes k[G]) \longrightarrow M \otimes k[G].$$

This determines a comodule structure of the form (1.11) by restricting to $\mathrm{Id}_{k[G]} \in \mathrm{Hom}_{k\text{-alg}}(k[G], k[G])$. Conversely, given a comodule structure $\Delta_M$, we get a pairing of the form (1.10) as the following composition

$$\mathrm{Hom}_{k\text{-alg}}(k[G], A) \times (M \otimes A) \longrightarrow \mathrm{Hom}_{k\text{-alg}}(k[G], A) \times (M \otimes k[G] \otimes A)$$

$$\longrightarrow M \otimes A \otimes A \longrightarrow M \otimes A$$

where the first map is given by $\Delta_M$, the second by the natural pairing, and the third by the ring structure on $A$.

If $M$ is a $G$-module, then the $G$-invariant submodule of $M$ is the $G$-submodule with trivial $G$-action given by

$$M^G = \{m \in M \mid g \cdot (m \otimes 1) = m \otimes 1, \ \forall \, A, g \in G(A)\}$$

which is readily seen to be equal

$$M^G = \{m \in M \mid \Delta_M(m) = m \otimes 1\}.$$

If $M, N$ are two $G$-modules, then the tensor product (over $k$) $M \otimes N$ has a natural structure of a $G$-module given by embedding $G$ diagonally in $G \times G$; this is written succinctly in terms of $\Delta_M, \Delta_N$ as the following composition involving the product structure of the ring $k[G]$:

$$\cdot \circ (\Delta_M \otimes \Delta_N) : M \otimes N \longrightarrow (M \otimes k[G]) \otimes (N \otimes k[G]) \longrightarrow M \otimes N \otimes k[G].$$

If the $G$-module $M$ is finite dimensional (as a $k$ vector space), we may give another useful formulation of the concept of a $G$-module. Namely, suppose that $M$ is $n$-dimensional and identify the affine group scheme of $k$-automorphisms of $M$ with $\mathrm{GL}_n$. Then to give $M$ the structure of a $G$-module is equivalent to giving a homomorphism $\rho_M : G \to \mathrm{GL}_n$ of affine group schemes.

An important example of a $G$-module is the coordinate algebra itself. We readily check that the coproduct on $k[G]$, $\Delta : k[G] \to k[G] \otimes k[G]$, corresponds to the right regular representation of $G$ on the functions of $G$: $(g \in G, f(-) \in k[G]) \mapsto f(- \cdot g) \in k[G]$.

Suppose that $H \subset G$ is a closed subgroup scheme of the affine group scheme $G$ (*i.e.*, $k[G] \to k[H]$ is surjective). Then for any $H$-module $N$, we consider the $H$-fixed points of $k[G] \otimes N$, where $H$ acts on $k[G]$ via the right regular representation. We use the notation

$$\mathrm{Ind}_{\mathrm{H}}^G N = (k[G] \otimes N)^H$$

to denote the $G$-module with $G$ action given by the left regular representation of $G$ on $k[G]$.

One very useful aspect of this induction functor is given by the following theorem which is often called *Frobenius reciprocity*.

**Theorem 1.12 (cf. [13], 3.4).** — *If $H \subset G$ is a closed subgroup of the affine group scheme $G$, then $\mathrm{Ind}_{\mathrm{H}}^G(-)$ is right adjoint to the restriction functor. In other words, for every $H$-module $N$ and every $G$-module $M$, there is a natural isomorphism*

$$\mathrm{Hom}_H(M, N) \simeq \mathrm{Hom}_G(M, \mathrm{Ind}_{\mathrm{H}}^G N).$$

*In particular, if $N$ is an injective $H$-module, then $\mathrm{Ind}_{\mathrm{H}}^G N$ is an injective $G$-module. For example, $k[G] = \mathrm{Ind}_e^G k$ is an injective $G$-module.*

Observe that sending $m \in M$ to $m \otimes \varepsilon \in M \otimes k[G]$ determines a homomorphism $M \to M \otimes k[G]$ of $G$-modules, where $\varepsilon : G \to k$ is evaluation at the identity (*i.e.*, the co-unit of the Hopf algebra $k[G]$). A direct calculation shows that the map $M \otimes k[G] \to M_{tr} \otimes k[G]$ defined by $m \otimes f \mapsto (1 \otimes f)\Delta_M(m)$ is an isomorphism of $G$-modules, where $M_{tr}$ is a trivial $G$-module isomorphic to $M$ as a $k$-vector space. Since $k[G]$ is an injective $G$-module, this verifies that any $G$-module can be embedded into an injective module.

Consequently, the category of $G$-modules is an abelian category with enough injectives, so that we may use standard homological algebra to define

$$\mathrm{Ext}_G^i(M, N) = R^i \mathrm{Hom}_G(M, -)(N)$$

for any pair of $G$-modules $M, N$. As usual, we denote $\text{Ext}_G^*(k, M)$ by $\text{H}^*(G, M)$, so that

$$\text{H}^i(G, M) = R^i \text{Hom}_G(k, -)(M) = R^i(-)^G(M)$$

where the $G$-fixed point functor $(-)^G$ sends a $G$-module $M$ to the $G$ fixed points of $M$, $M^G$, as defined above. We readily verify that Theorem 1.12 implies that

$$\text{H}^*(H, N) \simeq \text{H}^*(G, \text{Ind}_H^G N)$$

whenever $H \subset G$ is a closed subgroup scheme and $N$ is a $H$-module.

Let $\phi : k \to k$ denote the $p$-th power map which sends $\alpha \in k$ to $\alpha^p \in k$. ($\phi$ is often called the *arithmetic Frobenius map*.) Given a $k$ vector space $V$, we obtain a new $k$-vector space $V^{(1)}$ defined as the base change of $V$ via $\phi$,

$$V^{(1)} = k \otimes_\phi V.$$

If $k$ is perfect (*i.e.*, if $\phi$ is an isomorphism), then

$$V^{(1)} \simeq V, \quad \alpha \otimes v \longmapsto \alpha^{1/p} v$$

identifies $V^{(1)}$ via a semi-linear map with $V$, so that we may view $V^{(1)}$ as the vector space $V$ with the modified $k$-action given by $(\alpha, v) \mapsto \alpha^{1/p} v$. $V^{(1)}$ is called the (first) *Frobenius twist* of $V$.

***Definition 1.13***. — If $G$ is an affine group scheme, we denote by $G^{(r)}$ the affine group scheme whose coordinate algebra is $k[G]^{(r)}$, the $r$th Frobenius twist of $k[G]$. Moreover, we denote by $G_{(r)}$ the affine group scheme defined as the kernel of the natural map

$$G_{(r)} = \ker\{\Phi^r : G \longrightarrow G^{(r)}\},$$

where $\Phi^{r*} : k[G^{(r)}] \to k[G]$ is the $k$-linear map sending $f \in k[G]^{(r)}$ to $f^{p^r} \in k[G]$.

If $G$ is defined over the finite field $\mathbb{F}_{p^r}$ so that $G = G_{\mathbb{F}_{p^r}} \times_{\text{Spec}\, \mathbb{F}_{p^r}} \text{Spec}\, k$, then

$$\Phi^{r*} = F^{r*} \circ \phi^r : k[G]^{(r)} \simeq k[G] \longrightarrow k[G].$$

Here, $F^r$ is the so-called *geometric Frobenius* of $G$, defined as the base change from $\mathbb{F}_{p^r}$ to $k$ of the $p^r$-th power map on $\mathbb{F}_{p^r}[G_{\mathbb{F}_{p^r}}]$. Thus, for such $G$ we can identify $G_{(r)}$ with the kernel of $F^r$,

$$G_{(r)} = \ker\{F^r : G \longrightarrow G\}.$$

In the special case $G = \text{GL}_n$, we readily verify that $(\text{GL}_n)_{(r)}$ so defined equals $\text{GL}_{n(r)}$ as discussed in Example 1.5.

We conclude that whenever $G$ is defined over $\mathbb{F}_{p^r}$, a $G$-module $M$ determines a new $G$-module $M^{(r)}$, the $r$-th *Frobenius twist* of $M$. If $\rho_M : G \to \text{GL}_n$ is the representation associated to the $G$-module $M$, then

$$\rho_{M^{(r)}} = F^r \circ \rho : G \longrightarrow \text{GL}_n \longrightarrow \text{GL}_n$$

is the representation associated to $M^{(r)}$. Observe that $M^{(r)}$ is trivial as a $G_{(r)}$-module, so that

$$\mathrm{H}^0(G_{(r)}, M) \neq \mathrm{H}^0(G_{(r)}, M^{(r)})$$

whenever $M$ is non-trivial as a $G_{(r)}$-module. Similarly, the cohomology $\mathrm{H}^*(\mathrm{GL}_n, M)$ can be quite different from $\mathrm{H}^*(\mathrm{GL}_n, M^{(1)})$. Indeed, this difference plays an important role in our techniques for computation.

## 2. Cohomological techniques

Much of the second lecture of this series was dedicated to explaining weights associated to the action of a torus with the goal of giving some insight into the effect that Frobenius twist plays in cohomology. This written version adds to the original lecture by giving a brief introduction to some of the techniques used in the computation of cohomology. The reader is referred to the book of J. Jantzen [13] for a much more complete exposition of these techniques.

The algebraic group $\mathrm{GL}_1$ is typically denoted $\mathbb{G}_m$ and called the multiplicative group. The coordinate algebra $k[\mathbb{G}_m]$ is given by

$$k[\mathbb{G}_m] = k[t, t^{-1}] = k[u, v]/(uv - 1)$$

with coproduct $t \mapsto t \otimes t$. A split torus of rank $n$ is an algebraic group isomorphic to $\mathbb{G}_m^{\times n}$. The subgroup $T_n \subset \mathrm{GL}_n$ of diagonal matrices is the usual model for such a split torus of rank $n$.

The representation theory of a split torus is particularly easy to describe as the following proposition recalls.

**Proposition 2.1**. — *Every $T_n$-module splits as a direct sum of $1$-dimensional irreducible $T_n$-modules. An irreducible $T_n$-module is given by its weight $\underline{\lambda} = (\lambda_1, \ldots \lambda_n) \in \mathbb{Z}^{\oplus n}$, where for a given finitely generated $k$-algebra $A$ the diagonal matrix*

$$\begin{pmatrix} x_1 & & \\ & \ldots & \\ & & x_n \end{pmatrix} \in (A^*)^n$$

*acts on the rank $1$ $A$-module via multiplication by $x_1^{\lambda_1} \ldots x_n^{\lambda_n}$.*

*Similarly, every $T_{n(r)}$-module splits as a direct sum of $1$-dimensional $T_{n(r)}$-modules, where $T_n(r)$ is the $r$-th Frobenius kernel of $T_n$. The weights $\underline{\lambda} = (\lambda_1, \ldots \lambda_n)$ of $T_{n(r)}$ can be viewed as taking values in $\{0, 1, \ldots p^r - 1\}^n$ since any $(x_1, \ldots, x_n) \in T_{n(r)}(A)$ satisfies $x_i^{p^r} = 1$.*

**Example 2.2**. — The most basic example is the action of $T_n$ on an $n$-dimensional vector space given by multiplication; in this case, the weights of this action are all of the form $(0, \ldots, 0, 1, 0, \ldots 0)$. We view this action as given by the pairing of algebraic groups $\mu : T_n \times \mathbb{G}_a^{\times n} \to \mathbb{G}_a^{\times n}$, which is equivalent to the data of a compatible collection of pairings $\mu : (A^*)^{\times n} \times A^{\oplus n} \to A^{\oplus n}$ for every finitely generated $k$-algebra $A$.

A second basic example is the action of $T_n$ on $\mathrm{Hom}_{\mathrm{grp\ sch}}(\mathbb{G}_a^{\times n}, \mathbb{G}_a)$, the dual vector space. Since we define $\mathrm{Hom}_{\mathrm{grp\ sch}}(\mathbb{G}_a^{\times n}, \mathbb{G}_a)$ as a $\mathbb{G}_m$-module so that the evaluation pairing $\mathrm{Hom}_{\mathrm{grp\ sch}}(\mathbb{G}_a^{\times n}, \mathbb{G}_a) \times \mathbb{G}_a^{\times n} \to \mathbb{G}_a$ is $\mathbb{G}_m$ equivariant with $\mathbb{G}_m$ acting trivially on the right hand side, the resulting weights of $\mathrm{Hom}_{\mathrm{grp\ sch}}(\mathbb{G}_a^{\times n}, \mathbb{G}_a)$ are all of the form $(0, \dots, 0, -1, 0, \dots, 0)$. Observe that under this action

$$(\alpha_1, \dots, \alpha_n), (\psi_1(-), \dots \psi_n(-)) \longmapsto (\psi_1(\alpha_1^{-1} \cdot -), \dots, \psi_n(\alpha_n^{-1} \cdot -)),$$

which is the usual contragredient action.

Rather than discuss maximal tori and weights for general reductive groups, we describe the situation for the example of primary interest, that of the algebraic group $\mathrm{GL}_n$.

**Proposition 2.3**. — *Let $M$ be a $\mathrm{GL}_n$-module.*

(1) *As a $T_n$-module, $M \simeq \oplus M_{\underline{\lambda}}$. The $T_n$-submodule $M_{\underline{\lambda}} \subset M$ is called the $\underline{\lambda}$-weight subspace.*

(2) *The Frobenius twist $M^{(r)}$ of $M$ has weight decomposition $M^{(r)} = \oplus M_{p^r \underline{\lambda}}$.*

(3) *Let $\mathbb{G}_m \subset T_n$ denote the subgroup of scalar multiples of the identity. As a $\mathrm{GL}_n$-module, $M$ splits as a direct sum $M = \oplus M_d$, where $M_d \subset M$ is the weight subspace of weight $d$ with respect to the action of $\mathbb{G}_m$.*

Observe that $\mathrm{H}^i(T_n, M) = 0$, $i > 0$ since $T_n$ is semisimple. On the other hand, the cohomology of $\mathbb{G}_a$ is quite interesting. We recall its computation, including its weight structure where the action of $\mathbb{G}_m$ on $\mathrm{H}^*(\mathbb{G}_a, k)$ is that induced by the multiplication action of $\mathbb{G}_m$ on $\mathbb{G}_a$. Because the group algebra of $\mathbb{G}_{a(1)}$ is isomorphic to the group algebra of the finite group $\mathbb{Z}/p$, we know $\mathrm{H}^*(\mathbb{G}_{a(1)}, k)$ from our knowledge of the cohomology of finite cyclic groups.

**Theorem 2.4 (cf. [4])**

(1) $\mathrm{H}^*(\mathbb{G}_a, k) = \Lambda^*(y_1, y_2, \dots) \otimes k[x_1, x_2, \dots], \quad p \neq 2.$

$$\mathrm{H}^*(\mathbb{G}_a, k) = k[y_1, y_2, \dots], \quad p = 2$$

*where each $y_i \in \mathrm{H}^1(\mathbb{G}_a, k), x_i \in \mathrm{H}^2(\mathbb{G}_a, k)$.*

(2) *Let $F : \mathbb{G}_a \to \mathbb{G}_a$ be the (geometric) Frobenius endomorphism. Then*

$$F^*(x_i) = x_{i+1}, \quad F^*(y_i) = y_{i+1}.$$

(3) *The weight of $x_i$ is $-p^i$ and of $y_i$ is $-p^{i-1}$.*

(4) *If $(\alpha \cdot -) : \mathbb{G}_a \to \mathbb{G}_a$ denotes multiplication by $\alpha \in k$, then*

$$(\alpha \cdot -)^*(x_i) = \alpha^{p^i} x_i, \quad (\alpha \cdot -)^*(y_i) = \alpha^{p^{i-1}} y_i,$$

(5) $\mathrm{H}^*(\mathbb{G}_{a(r)}, k) = \Lambda^*(y_1, \ldots, y_r) \otimes k[x_1, \ldots, x_r], \quad p \neq 2.$

$$\mathrm{H}^*(\mathbb{G}_{a(r)}, k) = k[y_1, \ldots, y_r], \quad p = 2.$$

The reader puzzled about the fact that the generator $y_1 \in \mathrm{H}^1(\mathbb{G}_{a(1)}, k) = \mathrm{Hom}_{\mathrm{grp\ sch}}(\mathbb{G}_{a(1)}, \mathbb{G}_a)$ has weight $-1$ whereas the generator $x_1 \in \mathrm{H}^2(\mathbb{G}_{a(1)}, k)$ has weight $-p$ might find it helpful to know that $x_1$ is the Bockstein of $y_1$. Thus, if $y_1$ is represented by some function $f \in k[\mathbb{G}_{a(1)}]$, then $x_1$ is represented by $\delta(f) \in k[\mathbb{G}_{a(1)}^2]$ defined by

$$\delta(f)(g_1, g_2) = \frac{f(g_1^p) + f(g_2^p) - f(g_1^p + g_2^p)}{p}.$$

A very useful technique for computations is the Lyndon-Hochschild-Serre (L-H-S) (first quadrant, cohomological) spectral sequence

(2.5) $\qquad E_2^{p,q} = \mathrm{H}^p(G/N, \mathrm{H}^q(N, M)) \Longrightarrow \mathrm{H}^{p+q}(G, M)$

relating the cohomology of $G$ with coefficients in the $G$-module $M$ to the cohomology of $G/N$ with coefficients in the $G/N$-module $\mathrm{H}^*(N, M)$, the cohomology of the normal subgroup scheme $N$ with coefficients in the restriction of $M$ to $N$.

**_Example 2.6_**. — Let $B_n \subset \mathrm{GL}_n$ denote the subgroup of upper triangular matrices, and let $U_n \subset B_n$ denote the subgroup of strictly upper triangular matrices. We utilize the short exact sequence

$$1 \longrightarrow U_n \longrightarrow B_n \longrightarrow T_n \longrightarrow 1$$

and the semi-simplicity of $T_n$ to conclude that

$$\mathrm{H}^*(B_n, M) \simeq (\mathrm{H}^*(U_n, M))^{T_n}.$$

Similarly, for any $r \geqslant 1$, we conclude

$$\mathrm{H}^*(B_{n(r)}, M) \simeq (\mathrm{H}^*(U_{n(r)}, M))^{T_{n(r)}}.$$

In the special case of trivial coefficients (_i.e._, $M = k$), we may make further progress in the computation of $\mathrm{H}^*(B_n, k)$ by using a central series for $U_n$ to express $U_n$ as a succession of central extensions of products of root subgroups (_i.e._, subgroups isomorphic to $\mathbb{G}_a$ stabilized by $T_n$). Then the action of $T_n$ stabilizes each of these extensions and thus induces a $T_n$-action on their associated L-H-S spectral sequences.

Indeed, if we pass to the first Frobenius kernel $B_{n(1)}$ of $B_n$ and assume that $p > n$, then this strategy gives a complete calculation of $\mathrm{H}^*(B_{n(1)}, k)$ as a $T_{n(1)}$-module. Namely, we consider the height 1 central extensions

$$1 \longrightarrow \mathbb{G}_{a(1)} \longrightarrow U_{(1)} \longrightarrow \overline{U}_{(1)} \longrightarrow 1,$$

associated to this central series for $U_n$. Applying the exact functor $(-)^{T_{n(1)}}$ to each of the associated L-H-S spectral sequences, we obtain spectral sequences of the form

(2.7) $\qquad E_2^{p,q} = (\mathrm{H}^p(\overline{U}_{(1)}, k) \otimes \mathrm{H}^q(\mathbb{G}_{a(1)}, k))^{T_{n(1)}} \Longrightarrow \mathrm{H}^{p+q}(U_{(1)}, k)^{T_{n(1)}}.$

If $p > n$, then the computation of Theorem 2.4 together with the multiplicative structure of (2.5) implies that all of the differentials of (2.7) are 0. Thus, we obtain an isomorphism of $T_{(n)}$-modules

$$\mathrm{H}^*(U_{n(1)}, k) \simeq \mathrm{H}^*(\mathrm{gr}(U_n)_{(1)}, k), \quad \mathrm{gr}(U_n) \simeq \mathbb{G}_a^{\times N}, \ N = \frac{n(n-1)}{2}.$$

Assuming $p > n$, one can fully compute $\mathrm{H}^*(B_{n(1)}, k) = (\mathrm{H}^*(U_{n(1)}, k))^{T_{n(1)}}$ by taking the $T_{n(1)}$ invariants of

$$\mathrm{H}^*(\mathrm{gr}(U_n)_{(1)}, k) = \otimes_{i=1}^{N} \mathrm{H}^*(\mathbb{G}_{a(1)}, k).$$

Finally, a weight argument (for $p > n$) implies that $\mathrm{H}^*(B_{(1)}, k) \simeq S^*(u_n^{\#(1)})$ where $u_n = \mathrm{Lie}(U_n)$. (See [**13**, 12.12] for details of this weight argument; the earlier part of the above argument using the L-H-S spectral sequence is replaced in [**13**] by a different spectral sequence argument.)

As we recall in the following corollary of "Kempf's Vanishing Theorem", the co-homology $\mathrm{H}^*(\mathrm{GL}_n, M)$ is isomorphic to $\mathrm{H}^*(B_n, M)$. We state this theorem more generally for an arbitrary affine algebraic group $G$; we remind the reader that a Borel subgroup $B \subset G$ is a maximal closed, connected, reduced, solvable subgroup scheme.

**Theorem 2.8 (cf. [4], [18**, 3.1]). — *Let $G$ be an affine algebraic group and $B \subset G$ be a Borel subgroup. Then for any $G$-module $M$, the natural restriction map*

$$\mathrm{H}^*(G, M) \longrightarrow \mathrm{H}^*(B, M)$$

*is an isomorphism.*

**Example 2.9**. — A construction of G. Hochschild provides a natural map

$$g^{\#(1)} \longrightarrow \mathrm{H}^2(V(g), k),$$

where $g^\#$ is the linear dual of the $p$-restricted Lie algebra $g$. Namely, $\mathrm{H}^2(V(g), k)$ can be naturally identified with isomorphism classes of extensions of $p$-restricted Lie algebras of the form

$$1 \longrightarrow k \longrightarrow \tilde{g} \longrightarrow g \longrightarrow 1,$$

where $k$ is equipped with the trivial $p$-restriction as well as trivial Lie bracket. For any linear map $\psi : g \to k$, we define following Hochschild the $p$-restricted Lie algebra $\tilde{g}$ with Lie algebra structure the direct sum $k \oplus g$ and with $p$-restriction given by $(\alpha, X)^{[p]} = (\psi(X)^p, X^{[p]})$.

Let $G$ be an affine group scheme and let $I$ denote the augmentation ideal of $k[G]$, the maximal ideal at the identity $e \in G$. Then we set

$$\mathrm{gr}(k[G]) = \oplus_{n \geqslant 0} I^n / I^{n-1},$$

and readily verify that the commutative Hopf algebra structure on $k[G]$ determines a commutative Hopf algebra structure on $\mathrm{gr}(k[G])$. We denote the associated affine group scheme by $\mathrm{gr}(G)$. If $M$ is a $G$-module, then the standard "Hochschild complex"

$C^*(G, M)$ admits an associated filtration whose associated graded complex is the Hochschild complex $C^*(\mathrm{gr}(G), M)$ where $M$ is viewed as a trivial $\mathrm{gr}(G)$-module. This leads to the following general form of the "May spectral sequence."

**Theorem 2.10 (cf. [13], 9.13).** — *For any affine group scheme $G$ and $G$-module $M$, there is a natural first quadrant spectral sequence of cohomological type*

$$E_1^{s,t}(M) = \mathrm{H}^{s+t}(\mathrm{gr}(G)_{(s)} \otimes M) \Longrightarrow \mathrm{H}^{s+t}(G, M).$$

*For $G = \mathrm{GL}_{n(r)}$, this specializes to*

$$E_1^{*,*}(M) = \bigotimes_{i=1}^{r} S^*(\mathrm{gl}_n^{\#(i)}[2]) \otimes \Lambda^*(\mathrm{gl}_n^{\#(i-1)}[1]) \otimes M \Longrightarrow \mathrm{H}^*(\mathrm{GL}_{n(r)}, M),$$

*where $S^*(\mathrm{gl}_n^{\#(i)}[2])$ denotes the symmetric algebra generated by the vector space $\mathrm{gl}_n^{\#(i)}$ in degree 2, $\Lambda^*(\mathrm{gl}_n^{\#(i-1)}[1])$ the exterior algebra generated by $\mathrm{gl}_n^{\#(i-1)}$ in degree 1, and the notation specifies the structure of the spectral sequence with its $\mathrm{GL}_n$-action.*

**Example 2.11.** — We apply Example 2.9 and Theorem 2.10 to sketch a computation of $\mathrm{H}^*(V(\mathrm{gl}_n), k) = \mathrm{H}^*(\mathrm{GL}_{n(1)}, k)$ for $p \geqslant n$. Even though this sketch omits several somewhat difficult arguments, it can serve to suggest the manner in which computations can be made.

The May spectral sequence for $\mathrm{GL}_{n(1)}$ and $M = k$ has the form

$$E_1^{2s,t}(k) = S^s(\mathrm{gl}_n^{\#(1)}[2]) \otimes \Lambda^t(\mathrm{gl}_n, k) \Longrightarrow \mathrm{H}^{2s+t}(\mathrm{GL}_{n(1)}, k).$$

where $\mathrm{H}^t(\mathrm{gl}_n, k)$ is the Lie algebra cohomology of $\mathrm{gl}_n$ given as $\mathrm{H}^t(\Lambda^*(\mathrm{gl}_n^\#))$. The Hochschild construction of Example 2.9 implies that $E_1^{2,0}(k) = S^1(\mathrm{gl}_n^{\#(1)}[2])$ consists of permanent cycles; by multiplicativity of the spectral sequence, we conclude that $E_1^{*,0}(k) = S^*(\mathrm{gl}_n^{\#(1)}[2])$ consists of permanent cycles. A direct computation of $d_1^{0,*}$ implies that $E_2^{0,*}(k) = \mathrm{H}^*(\mathrm{gl}_n, k)$, the cohomology of the universal enveloping algebra of the Lie algebra $\mathrm{gl}_n$. Thus, the $E_2$-page of the May spectal sequence has the form

$$E_2^{2s,t}(k) = S^s(\mathrm{gl}_n^{\#(1)}[2]) \otimes \mathrm{H}^t(\mathrm{gl}_n, k) \Longrightarrow \mathrm{H}^{2s+t}(\mathrm{GL}_{n(1)}, k).$$

As verified in [9, 1.1] if $p > n$ then

$$\mathrm{H}^*(\mathrm{gl}_n, k) = (\Lambda^*(\mathrm{gl}_n^\#))^{(\mathrm{GL}_n)_1}$$

is an exterior algebra on generators in degrees $1, 3, \ldots, 2n - 1$, whereas the latter is shown in [1] (*cf.* [13, 12.10]) to be isomorphic to $(\Lambda^*(\mathrm{gl}_n^\#))^{\mathrm{GL}_n}$. We assume inductively that the first $i$ generators of $\mathrm{H}^*(\mathrm{gl}_n, k)$ transgress to some non-zero element of $E_{2i}^{*,0}(k)$. An argument of Borel enables us to conclude that on the $E_{2i+1}$-page we have $E_{2i+1}^{*,j}(k) = 0, 0 < j \leqslant i$, so that $E_{2i+1}^{*,0}(k)$ is the quotient of $S^*(\mathrm{gl}_n^{\#(1)}[2])$ by the ideal generated by the transgressions of elements of $\oplus_{j=1}^{i} \mathrm{H}^j(\mathrm{gl}_n, k)$.

On the other hand, $\mathrm{H}^{*>0}(\mathrm{GL}_n, k) = 0$; this is a special case of the usual Kempf vanishing theorem, but could be rederived using Theorem 2.8 and a weight argument showing $\mathrm{H}^{*>0}(U_n, k)^{T_n} = 0$. The $\mathrm{GL}_n$ invariance of $\mathrm{H}^*(\mathrm{gl}_n, k)$ implies that $i + 1$-st generator of $\mathrm{H}^*(\mathrm{gl}_n, k)$ must transgress to some non-zero element of $E^{2i,0}_{2i+2}(k)$. We conclude that $\mathrm{H}^*(\mathrm{GL}_{n(1)}, k)$ is isomorphic to $S^*(\mathrm{gl}_n^{(1)\#}[2])$ modulo the ideal generated by the transgressions of $\mathrm{H}^{*>0}(\mathrm{gl}_n, k)$ (which necessarily equals the ideal generated by the $\mathrm{GL}_n^{(1)}$-invariant elements of positive degree). (See [**10**] for details.)

We conclude this lecture by examining the fundamental class

$$(2.12) \qquad\qquad e_1 \in \mathrm{H}^2(\mathrm{GL}_n, \mathrm{gl}_n^{(1)}) \simeq \mathrm{Ext}^2_{\mathrm{GL}_n}(V_n^{(1)}, V_n^{(1)})$$

where $\mathrm{gl}_n$ denotes the adjoint module (*i.e.*, $n^2$-dimensional vector space of $n \times n$ matrices with $\mathrm{GL}_n$ acting via conjugation) and $V_n$ is the natural $\mathrm{GL}_n$-module associated to the identity representation $\mathrm{GL}_n \to \mathrm{Aut}_k(V_n)$. This fundamental class enables a straight-forward proof of the finite generation of $\mathrm{H}^*(\mathrm{GL}_{n(1)}, k)$ (*cf.* Theorem 4.1). The role of strict polynomial functors and their cohomology in Lecture 4 will be to establish suitable higher order fundamental classes $e_r$ which will enable the proof of finite generation of $\mathrm{H}^*((\mathrm{GL}_{n(r)}, k)$ and thus the cohomology of any finite group scheme.

First, observe that

$$\mathrm{H}^2(\mathrm{GL}_n, \mathrm{gl}_n) = \mathrm{H}^2(B_n, \mathrm{gl}_n) = 0 = \mathrm{H}^2(U_n, \mathrm{gl}_n)^{T_n} = 0$$

because no weight of $\mathrm{H}^2(U_n, k)$ is the negative of a weight of $\mathrm{gl}_n$ provided that $p > n$. (This can be verified as in Example 2.6 using the computation of Theorem 2.4 or more directly using the May spectral sequence.) This emphasizes the role of Frobenius twists. The possibility that $\mathrm{H}^2(\mathrm{GL}_n, \mathrm{gl}_n^{(1)})$ is non-zero can be seen from our knowledge that the weights of $\mathrm{gl}_n^{(1)}$ are $p$ times the weights of $\mathrm{gl}_n$.

***Example 2.13***. — Let $W_2(k)$ denote the Witt vectors of length 2 over $k$, so that $W_2(k)$ is the Artinian $k$-algebra whose underlying additive structure is as a non-trivial extension of $k$ by $k$. (Thus, $W_2(\mathbb{F}_p) = \mathbb{Z}/p^2\mathbb{Z}$.) Then we have an extension of affine group schemes over $k$,

$$1 \longrightarrow \mathrm{gl}_n^{(1)} \longrightarrow \mathrm{GL}_{n, W_2(k)} \longrightarrow \mathrm{GL}_n \longrightarrow 1$$

which corresponds to a class in $\mathrm{H}^2(\mathrm{GL}_n, \mathrm{gl}_n^{(1)})$; since this extension does not split, this class is non-trivial and is one representation of our fundamental class $e_1$.

Another representation of the class $e_1$ uses the May spectral sequence of Theorem 2.10 for $G = \mathrm{GL}_{n(1)}$ and $M = \mathrm{gl}_n^{(1)}$. There is a canonical $\mathrm{GL}_n^{(1)}$-invariant "identity element" $\mathrm{Id} \in \mathrm{gl}_n^{(1)} \otimes \mathrm{gl}_n^{\#(1)} \simeq (E_2^{2,0})(\mathrm{gl}_n^{(1)})$ which determines a class in $\mathrm{H}^2(\mathrm{GL}_n, \mathrm{gl}_n^{(1)})$ using the L-H-S spectral sequence for the short exact sequence

$$1 \longrightarrow \mathrm{GL}_{n(1)} \longrightarrow \mathrm{GL}_n \longrightarrow \mathrm{GL}_n^{(1)} \longrightarrow 1.$$

As yet another representation of $e_1$, we consider the exact sequence of $\mathrm{GL}_n$-modules

$$(2.14) \qquad\qquad 0 \longrightarrow V_n^{(1)} \longrightarrow S^p(V_n) \longrightarrow \Gamma^p(V_n) \longrightarrow V_n^{(1)} \longrightarrow 0,$$

where $S^p(V_n) = (V_n^{\otimes p})/\mathfrak{S}_p$ is the $p$-th symmetric power of $V_n$ represented concretely by the vector space of polynomials in $n$ variables homogeneous of degree $p$, and

$\Gamma^p(V_n) = (V_n^{\otimes p})^{\mathfrak{S}_p} = (S^p(V_n))^{\#}$. The map $V_n^{(1)} \to S^p(V_n)$ is given by $v \mapsto v^p$, the map $\Gamma^p(V_n) \to V_n^{(1)}$ is the dual of this map, and the map $S^p(V_n) \to \Gamma^p(V_n)$ is the symmetrization map. The extension (2.14) corresponds to the class

$$e_1 \in \operatorname{Ext}^2_{\operatorname{GL}_n}(V_n^{(1)}, V_n^{(1)}) \simeq \operatorname{H}^2(\operatorname{GL}_n, \mathfrak{gl}_n^{(1)}).$$

## 3. Polynomial modules and functors

In this lecture, we restrict our attention to $\operatorname{GL}_n$. A $\operatorname{GL}_n$-module is frequently called a *rational representation*, for the data necessary to provide an $N$-dimensional vector space with the structure of a $\operatorname{GL}_n$-module consists of $N^2$-matrix coefficients viewed as regular functions on $\operatorname{GL}_n$. Regular functions on $\operatorname{GL}_n$ can in turn be viewed as rational functions in the $n^2$ matrix coordinates of $\operatorname{GL}_n$. Should these $N^2$ rational functions all be polynomial functions of the matrix coordinates of $\operatorname{GL}_n$, namely lie in

$$(3.1) \qquad k[M_n] = k[x_{i,j}]_{1 \leqslant i,j \leqslant n} \subset k[x_{i,j}; t]_{1 \leqslant i,j \leqslant n} / \det(x_{i,j})t - 1 = k[\operatorname{GL}_n],$$

then the $\operatorname{GL}_n$-module is said to be a *polynomial module* (or a polynomial representation of $\operatorname{GL}_n$).

In this lecture, we shall see how to interpret such polynomial modules and their cohomology in terms of "strict polynomial functors" and we shall see how this functor point of view affords computational advantages. The formulation of strict polynomial functors is at first somewhat daunting, but the reader should keep in mind the fact that these functors are so defined in order to play the same role in connection with polynomial representations of $\operatorname{GL}_n$ as the role played by more familiar polynomial functors in connection with representations of the discrete group $\operatorname{GL}_n(k)$.

We begin with the definition of the *Schur algebra*.

***Definition 3.2***. — Let $n, d$ be position integers, consider the Hopf algebra $k[M_n]$ of (3.1), and let $k[M_n]_d \subset k[M_n]$ denote the subspace of homogeneous polynomials of degree $d$. Then $k[M_n]_d$ is closed under the coproduct of $k[M_n]$ and its linear dual (which is a finite dimensional $k$-algebra)

$$S(n, d) = (k[M_n]_d)^{\#}$$

is called the Schur algebra (of rank $n$ and degree $d$).

A module for $S(n, d)$ is called a polynomial module for $\operatorname{GL}_n$ homogeneous of degree $d$.

Thus, a module for $S(n, d)$ is a comodule for $k[M_n]_d \subset k[\operatorname{GL}_n]$, a $\operatorname{GL}_n$-module whose matrix coefficients are homogenous polynomial functions of the matrix coordinates of $\operatorname{GL}_n$. For future reference, we recall that

$$(3.3) \qquad S(n, d) = (S^d(\operatorname{End}_k(k^n))^{\#} = \Gamma^d(\operatorname{End}_k(k^n))$$
$$= ((\operatorname{End}_k(k^n))^{\otimes d})^{\mathfrak{S}_d} = \operatorname{End}_{k\mathfrak{S}_d}((k^n)^{\otimes d})$$

Let $\mathcal{P}ol_{n,d} \subset (\text{Mod}_{\text{GL}_n})$ denote the full-subcategory of polynomial modules for $\text{GL}_n$ homogeneous of degree $d$. Then essentially by definition we have an equivalence of categories

$$(3.4) \qquad \mathcal{P}ol_{n,d} \simeq (\text{Mod}_{S(n,d)})$$

between this category and the category of modules for the Schur algebra $S(n,d)$.

The following theorem tells us that $\text{GL}_n$-cohomology of polynomial modules can be computed as the cohomology of Schur algebras.

**Theorem 3.5 (cf. [5], [12, 3.12.1]).** — *Let*

$$\mathcal{P}ol_n = \oplus_{d \geqslant 0} \mathcal{P}ol_{n,d} \subset (\text{Mod}_{\text{GL}_n})$$

*denote the category of polynomial modules for* $\text{GL}_n$.

*(1) A* $\text{GL}_n$-*module is polynomial if and only if all of its* $T_n$-*weights are non-negative.*

*(2) Any polynomial module $M$ for $\text{GL}_n$ is canonically a direct sum of polynomial modules for $\text{GL}_n$ homogeneous of degree $d$, $M = \oplus_{d \geqslant 0} M_d$. Moreover, the homogenous summand of degree $d$ is the weight space of degree $d$ of the $\text{GL}_n$-module with respect to the scalar matrices $\mathbb{G}_m \subset T_n$.*

*(3) If $M, N$ are polynomial modules for $\text{GL}_n$, then*

$$\text{Ext}^*_{\text{GL}_n}(M, N) \simeq \text{Ext}^*_{\mathcal{P}ol_n}(M, N).$$

Let $\mathcal{V}_k$ denote the category of vector spaces over $k$ and let $(\mathcal{V}_k)^f \subset \mathcal{V}_k$ denote the full subcategory of finite dimensional $k$-vector spaces. As defined below, a strict polynomial functor is a collection of polynomial modules $M_n$ for $\text{GL}_n$ for each $n \geqslant 1$ together with compatibility of actions as $n$ varies.

**Definition 3.6.** — A strict polynomial functor $T : (\mathcal{V}_k)^f \to (\mathcal{V}_k)^f$ is the data of an association

$$T(V) \in (\mathcal{V}_k)^f, \quad \forall V \in (\mathcal{V}_k)^f$$

together with maps of affine schemes

$$T_{V,W} : \text{Hom}_k(V, W) \longrightarrow \text{Hom}_k(T(V), T(W)), \quad \forall V, W \in (\mathcal{V}_k)^f$$

satisfying the following:

- $T_{V,V}(\text{Id}_V) = \text{Id}_{T(V)}, \forall V \in (\mathcal{V}_k)^f.$
- $\forall U, V, W \in (\mathcal{V}_k)^f,$

$$
\begin{array}{ccc}
\text{Hom}_k(U,V) \times \text{Hom}_k(V,W) & \longrightarrow & \text{Hom}_k(U,W) \\
{\scriptstyle T_{U,V} \times T_{V,W}} \downarrow & & \downarrow {\scriptstyle T_{U,W}} \\
\text{Hom}_k(T(U),T(V)) \times \text{Hom}_k(T(V),T(W)) & \longrightarrow & \text{Hom}_k(T(U),T(W))
\end{array}
$$

commutes, where the horizontal maps are given by composition.

If $T$ is a strict polynomial functor with the property that $T_{V,W}$ has degree bounded by some integer which can be chosen independent of $V, W$, then we say that $T$ has bounded degree; if each $T_{V,W}$ is homogeneous of degree $d$, then we say that $T$ is homogeneous of degree $d$.

We denote by $\mathcal{P}_d$ the category of strict polynomial functors homogeneous of degree $d$, and $\mathcal{P}$ the category of strict polynomial functors of bounded degree.

**Remark 3.7**. — If the field $k$ is infinite, then a strict polynomial functor $T$ can be described more simply as a functor $T : (\mathcal{V}_k)^f \to (\mathcal{V}_k)^f$ with the property that each $T_{V,W} : \mathrm{Hom}_k(V, W) \to \mathrm{Hom}_k(T(V), T(W))$ is a polynomial function (*i.e.*, a map of sets having the property that with respect to a choice of bases for $\mathrm{Hom}_k(V, W), \mathrm{Hom}_k(T(V), T(W))$ the coordinates of $T_{V,W}(f)$ are polynomial in the coordinates of $f \in \mathrm{Hom}_k(V, W)$).

To compare with the former chapter's definition (Section 4.1), observe that a map of affine schemes
$$F : \mathrm{Hom}_k(V, W) \longrightarrow \mathrm{Hom}_k(T(V), T(W))$$
is equivalent to a map of coordinate algebras
$$F^* : S^*(\mathrm{Hom}_k(T(V), T(W))^{\#}) \longrightarrow S^*(\mathrm{Hom}_k(V, W)^{\#})$$
which is equivalent to a linear map of $k$-vector spaces
$$\mathrm{Hom}_k(T(V), T(W))^{\#} \longrightarrow S^*(\mathrm{Hom}_k(V, W)^{\#}).$$
To say that $F$ is homogeneous of degree $d$ is to say that this last map has image in $S^d(\mathrm{Hom}_k(V, W)^{\#}) \subset S^*(\mathrm{Hom}_k(V, W)^{\#})$, so that the data associated to this map is equivalent to a linear map of the form $\Gamma^d(\mathrm{Hom}_k(V, W)) \to \mathrm{Hom}_k(T(V)), T(W))$.

Thus, if $T$ is a strict polynomial functor homogeneous of degree $d$, we may replace the structure maps $T_{V,W}$ by equivalent linear maps which we shall continue to denote by $T_{V,W}$,
$$T_{V,W} : \Gamma^d(\mathrm{Hom}_k(V, W)) \longrightarrow \mathrm{Hom}_k(T(V)), T(W)).$$

**Proposition 3.8**. — *Let $T$ be a strict polynomial functor homogenous of degree $d$. Then $T(k^n)$ has the natural structure of a polynomial module for $\mathrm{GL}_n$ for each $n \geqslant 0$,*

*Proof.* — If $A$ is any finitely generated commutative $k$-algebra, then we may define $T(A \otimes V)$ to be $A \otimes T(V)$ and we may consider the base-change of $T_{V,W}$ to obtain
$$T_{V,W} \otimes A : \mathrm{Hom}_A(A \otimes V, A \otimes W) \longrightarrow \mathrm{Hom}_A(A \otimes T(V), A \otimes T(W)).$$
In particular, the composition
$$\mathrm{GL}_n(A) \subset \mathrm{Hom}_A(A \otimes k^n, A \otimes k^n) \longrightarrow \mathrm{Hom}_A(A \otimes T(k^n), A \otimes T(k^n))$$
for varying $A$ determines a $\mathrm{GL}_n$-module structure on $T(k^n)$ which by construction is polynomial. $\square$

**Example 3.9.** — We give some common examples of strict polynomial functors.

(1) The identity $I : (\mathcal{V}_k)^f \to (\mathcal{V}_k)^f$ is a strict polynomial functor homogeneous of degree 1.

(2) For any $r \geqslant 1$, $I^{(r)} : (\mathcal{V}_k)^f \to (\mathcal{V}_k)^f$ given by $V \mapsto V^{(r)}$ is a strict polynomial functor homogeneous of degree $p^r$.

(3) For any $d > 0$, $\otimes^d : (\mathcal{V}_k)^f \to (\mathcal{V}_k)^f$ given by $V \mapsto V^{\otimes^d}$ is a strict polynomial functor homogeneous of degree $d$.

(4) For any $d > 0$, $\Lambda^d : (\mathcal{V}_k)^f \to (\mathcal{V}_k)^f$ given by $V \mapsto \Lambda^d(V)$ is a strict polynomial functor homogeneous of degree $d$.

(5) For any $d > 0$, $S^d : (\mathcal{V}_k)^f \to (\mathcal{V}_k)^f$ given by $V \mapsto S^d(V)$ is a strict polynomial functor homogeneous of degree $d$.

(6) For any $d > 0$, $\Gamma^d : (\mathcal{V}_k)^f \to (\mathcal{V}_k)^f$ given by $V \mapsto \Gamma^d(V)$ is a strict polynomial functor homogeneous of degree $d$. More generally, for any $n, d > 0$, $\Gamma^d(\mathrm{Hom}_k(k^n, -))$ is a strict polynomial functor of degree $d$.

(7) If $T : (\mathcal{V}_k)^f \to (\mathcal{V}_k)^f$ is a strict polynomial of degree $d$, then $T^{\#}$ given by $V \mapsto T(V^{\#})^{\#}$ is also a strict polynomial functor of degree $d$. Moreover, $T$ is a projective object of the category $\mathcal{P}$ of strict polynomial functors of bounded degree if and only if $T^{\#}$ is an injective object of $\mathcal{P}$.

The following proposition makes more explicit various homological algebra constructions in $\mathcal{P}$, our category of strict polynomial functors of bounded degree. Observe that if $T$ is a strict polynomial functor homogeneous of degree $d$ then there is a natural map (*i.e.*, natural transformation of functors)

(3.10) $$T(k^n) \otimes \Gamma^d(\mathrm{Hom}_k(k^n, -)) \longrightarrow T$$

given for each $V \in (\mathcal{V}_k)^f$ as the adjoint of

$$T_{k^n, V} : \Gamma^d(\mathrm{Hom}_k(k^n, V)) \longrightarrow \mathrm{Hom}_k(T(k^n), T(V)).$$

**Proposition 3.11 (*cf.* [12, 2.10]).** — *The category $\mathcal{P}$ of strict polynomial functors of bounded degree is isomorphic to the direct sum of categories strict polynomial functors homogeneous of degree $d$ for $d > 0$,*

$$\mathcal{P} \simeq \oplus_d \mathcal{P}_d.$$

*Moreover, for any $n > 0$, the functor $\Gamma^d(\mathrm{Hom}_k(k^n, -)) \in \mathcal{P}_d$ is a projective object.*

*If $T$ is a strict polynomial functor homogeneous of degree $d$, then the natural map (3.10) is surjective provided that $n \geqslant d$. Thus, for $n \geqslant d$, $\Gamma^d(\mathrm{Hom}_k(k^n, -))$ is a projective generator of $\mathcal{P}_d$.*

We now formulate the theorem that tells us that we can compute Ext-groups of polynomial $\mathrm{GL}_n$-modules in terms of Ext-groups of strict polynomial functors.

**Theorem 3.12** (**[12**, 3.2]). — *For positive integers $n \geqslant d$, there are natural equivalences of abelian categories (with enough injective and projective objects)*

$$\mathcal{P}_d \simeq (\mathrm{Mod}_{S(n,d)}) \simeq \mathcal{P}ol_{n,d} \,.$$

*Consequently, for any pair of strict polynomial functors $S, T$ homogeneous of degree $d$, there are natural isomorphisms of graded groups*

$$(3.13) \qquad \mathrm{Ext}^*_{\mathcal{P}_d}(S,T) \simeq \mathrm{Ext}^*_{\mathcal{P}ol_{n,d}}(S(k^n), T(k^n))$$

$$\simeq \mathrm{Ext}^*_{S(n,d)}(S(k^n), T(k^n)) \simeq \mathrm{Ext}^*_{\mathrm{GL}_n}(S(k^n), T(k^n)).$$

*Outline of proof.* — The map $\mathcal{P}_d \to (\mathrm{Mod}_{S(n,d)})$ is given by $T \mapsto T(k^n)$. The action of $S(n,d) = \Gamma^d(\mathrm{End}_k(k^n))$ (*cf.* (3.3)) on $T(k^n)$ is given by (3.10). The proof that this is an equivalence of categories is more or less a direct computation using the explicit inverse sending $(\mathrm{Mod}_{S(n,d)}) \to \mathcal{P}_d$ given by $M \mapsto \Gamma^d(\mathrm{Hom}(k^n, -)) \otimes M$. The equivalence $(\mathrm{Mod}_{S(n,d)}) \simeq \mathcal{P}ol_{n,d}$ is that of 3.4.

The first three isomorphisms of (3.13) follow from the equivalences of categories. The last is given by (3.5). $\qquad\square$

We conclude this lecture by mentioning a few of the computational advantages one has when computing $\mathrm{Ext}_{\mathcal{P}}$-groups. One is the existence of *complexes of functors* (discussed in some detail in other lectures). For example, one has the (exact) Koszul complex

$$(3.14) \qquad 0 \longrightarrow \Lambda^d \longrightarrow \Lambda^{d-1} \otimes S^1 \longrightarrow \cdots \longrightarrow \Lambda^1 \otimes S^{d-1} \longrightarrow S^d \longrightarrow 0$$

and the not necessarily exact de Rham complex

$$(3.15) \qquad 0 \longrightarrow S^d \longrightarrow \Lambda^1 \otimes S^{d-1} \longrightarrow \cdots \longrightarrow \Lambda^{d-1} \otimes S^1 \longrightarrow \Lambda^d \longrightarrow 0$$

A second advantage is the very concrete nature of injectives and projectives. For example, the functors $S^d$ are injective and thus cohomologically acyclic.

A third useful computational tool, especially in conjunction with the above complexes, is the following acyclicity result originally due to T. Pirashvili **[14]**. This result appears in much strengthened form in **[7**, 1.7].

**Theorem 3.16** (*cf.* **[8]**, **[12**, 2.13]). — *Let $T, T'$ be homogenous strict polynomial functors of positive degree and let $A$ be an additive functor (e.g., a Frobenius twist of a strict polynomial functor homogeneous of degree 1). Then*

$$\mathrm{Ext}^*_{\mathcal{P}}(A, T \otimes T') = 0.$$

## 4. Finite generation of cohomology

In this lecture, we outline the proof of the finite generation of the cohomology of finite group schemes, Theorem 4.7. We first sketch the proof of finite generation for infinitesimal group schemes of height 1, for this special case introduces the general method of proving finite generation. We then discuss the existence and basic properties of the fundamental classes

$$e_r \in \mathrm{Ext}_{\mathcal{P}}^{2p^{r-1}}(I^{(r)}, I^{(r)}).$$

Using these classes, we then sketch the proof of finite generation of $\mathrm{H}^*(\mathrm{GL}_{n(r)}, k)$. Finally, we discuss the relatively straight-forward manner in which finite generation of $\mathrm{H}^*(\mathrm{GL}_{n(r)}, k)$ implies the finite generation of $\mathrm{H}^*(G, k)$ for any finite group scheme $G$.

The following theorem was first formulated and proved in [10] although the result might well have been known previously.

***Theorem 4.1*** ([10]). — *Let $G$ be an infinitesimal group scheme of height 1 (i.e., $k[G]$ is a finite connected algebra whose maximal ideal consists of elements whose p-th power is 0) and let $M$ be a finite dimensional G-module. Then $\mathrm{H}^*(G, k)$ is a finitely generated algebra and $\mathrm{H}^*(G, M)$ is a finite module over $\mathrm{H}^*(G, k)$.*

*Sketch of proof.* — As in Example (2.11), the Hochschild construction of Example 2.9 implies that the May spectral sequence of Theorem 2.10 has the form

$$E_2^{2s,t}(M) = S^s(g^{(1)\#}) \otimes \mathrm{H}^t(g, M) \implies \mathrm{H}^{2s+t}(G, M).$$

Here $\mathrm{H}^*(g, M)$ is the Lie algebra cohomology of $g = \mathrm{Lie}(G)$ (*i.e.*, the cohomology of the universal enveloping algebra $U(g)$ of $g$). The "shape" of this spectral sequence implies that $S^*(g^{(1)\#}) = E_2^{*,0}(k)$ consists of "permanent cycles" (*i.e.*, the differentials $d_r$ vanish on $E_r^{*,0}(k)$). This implies that $E_r^{*,*}(M)$ is a module over $S^*(g^{(1)\#})$ and that $d_r$ is a homomorphism of $S^*(g^{(1)\#})$-modules.

Now assume that $M$ is finite dimensional. Then $E_2^{*,*}(M)$ is a finite $S^*(g^{(1)\#})$-module. Since $E_r^{*,*}(M)$ is a subquotient of $E_{r-1}^{*,*}(M)$, we conclude that each $E_r^{*,*}(M)$ and thus also $E_\infty^{*,*}(M)$ are finite $S^*(g^{(1)\#})$-modules. In particular, $E_\infty^{*,*}(k)$ is a finite $S^*(g^{(1)\#})$-modules which implies that $E_\infty^{*,*}(k)$ is finitely generated which implies by a result of L. Evens [6, 2.1] that $\mathrm{H}^*(G, k)$ is finitely generated. Moreover, the spectral sequence $\{E_r^{*,*}(M)\}$ is a module over the spectral sequence (of algebras) $\{E_r^{*,*}(k)\}$, so that the action of $S^*(g^{(1)\#})$ on $E_r^{*,*}(M)$ factors through $E_r^{*,*}(k)$ and thus the action of $S^*(g^{(1)\#})$ on $\mathrm{H}^*(G, M)$ factors through $\mathrm{H}^*(G, k)$. We therefore conclude that $\mathrm{H}^*(G, M)$ is a finite $\mathrm{H}^*(G, k)$-module. $\qquad\square$

To extend this argument to more general finite group schemes $G$, we require a finitely generated subalgebra of $E_2^{*,*}(k)$ for the May spectral sequence consisting of permanent cycles and with respect to which the $E_r^{*,*}(M)$ is a finite module. We can no longer argue that the shape of May spectral sequence guarantees the existence of such an algebra. Instead, we construct explicit generators of such an algebra, the fundamental classes $\{e_i\}$, whose basic properties suffice to guarantee that they generate such an algebra.

The following complete calculation of $\mathrm{Ext}^*_{\mathcal{P}}(I^{(r)}, I^{(r)})$ is the heart of the proof of finite generation. The proof follows closely the arguments of [**8**].

***Theorem 4.2*** ([**12**, 4.10]). — *The* Ext*-algebra* $\mathrm{Ext}^*_{\mathcal{P}}(I^{(r)}, I^{(r)})$ *is a commutative $k$-algebra generated by elements*

$$e_i^{(r-i)} \in \mathrm{Ext}^{2p^{i-1}}_{\mathcal{P}}(I^{(r)}, I^{(r)})$$

*subject only to the relations* $(e_i^{(r-i)})^p = 0$.

*Comments on proof.* — We use all of the computational tools mentioned at the end of Lecture 3. Namely, we proceed by induction first on $r$ and then for a given $r$ by induction on $j$ to compute $\mathrm{Ext}^*_{\mathcal{P}}(I^{(r)}, S^{p^{r-j}}(j))$. Inputs to this computation include the vanishing of $\mathrm{Ext}^{*>0}_{\mathcal{P}}(-, S^d)$ for any $d \geqslant 0$ because of the injectivity of $S^d$, the vanishing of $\mathrm{Ext}^*_{\mathcal{P}}(I^{(r)}, \Lambda^i \otimes S^j)$ for $i, j > 0$ by Theorem 3.16, and the exactness of the Koszul complex (3.14). One additional input which enables this computation is a theorem of P. Cartier [**3**] which determines the cohomology of the de Rham complex (3.15); namely, the de Rham complex is acyclic if $(p, d) = 1$ and equals the first Frobenius twist of the de Rham complex relating $S^{d/p}$ to $\Lambda^{d/p}$ if $p \mid d$.          $\square$

Further work with Ext-groups in the category $\mathcal{P}$ of strict polynomial functors of bounded degree verifies that $e_r$ is related in a natural way to a power of $e_1$.

***Theorem 4.3*** ([**12**, 5.7]). — *The image of*

$$(e_1^{p-1})^{\otimes p^{r-1}} \in (\mathrm{Ext}^{2(p-1)}_{\mathcal{P}})(I^{(1)}, I^{(1)}))^{\otimes p^{r-1}}$$

*is a scalar multiple of the image of*

$$e_r^{p-1} \in \mathrm{Ext}^{2(p-1)p^{r-1}}_{\mathcal{P}}(I^{(r)}, I^{(r)})$$

*in* $\mathrm{Ext}^{2(p-1)p^{r-1}}_{\mathcal{P}}(\Gamma^{p^{r-1}}(1), S^{p^{r-1}}(1))$.

Theorem 4.3 enables us to conclude the existence of non-zero classes in the cohomology of $\mathrm{GL}_n$ which restrict non-trivially to the cohomology of $\mathrm{GL}_{n(1)}$.

***Theorem 4.4*** ([**12**, 6.2]). — *For any $n \geqslant 2, r \geqslant 1$, the image of $e_r$ under the composition*

$$\mathrm{Ext}^{2p^{r-1}}_{\mathcal{P}}(I^{(r)}, I^{(r)}) \longrightarrow \mathrm{Ext}^{2p^{r-1}}_{\mathrm{GL}_n}(V_n^{(r)}, V_n^{(r)})$$

$$= \mathrm{H}^{2p^{r-1}}(\mathrm{GL}_n, \mathfrak{gl}_n^{(r)}) \longrightarrow \mathrm{H}^{2p^{r-1}}(\mathrm{GL}_{n(1)}, k) \otimes \mathfrak{gl}_n^{(r)}$$

*is non-zero.*

Theorem 4.4 together with a bit more work implies the following corollary.

**Corollary 4.5**. — *The class* $e_r \in \mathrm{H}^{2p^{r-1}}(\mathrm{GL}_n, \mathrm{gl}_n^{(r)})$ *restricts to a non-trivial class in* $\mathrm{H}^{2p^{r-1}}(\mathrm{GL}_{n(r)}, k) \otimes \mathrm{gl}_n^{(r)}$. *We view this restriction as a non-zero map*

$$e_r : \mathrm{gl}_n^{(r)\#} \longrightarrow \mathrm{H}^{2p^{r-1}}(\mathrm{GL}_{n(r)}, k).$$

*This map annihilates the 1-dimensional* $\mathrm{GL}_n$-*invariant subspace of* $\mathrm{gl}_n^{(r)\#}$. *Moreover, its composition with the restriction map to* $\mathrm{H}^{2p^{r-1}}(\mathrm{GL}_{n(1)}, k)$ *is given up to non-zero scalar multiple as the composition*

$$\mathrm{gl}_n^{(r)\#} \longrightarrow S^{p^{r-1}}(\mathrm{gl}_n^{(1)\#}[2]) \longrightarrow \mathrm{H}^{2p^{r-1}}(\mathrm{GL}_{n(1)}, k)$$

*where the first map is the* $p^{r-1}$-*st power map and the second is the edge homomorphism in the May spectral sequence of Theorem 2.10.*

Corollary 4.5 enables us to adapt the proof of Theorem 4.1 to provide a proof of finite generation of cohomology of $\mathrm{GL}_{n(r)}$.

**Theorem 4.6**. — *Let* $n \geqslant 2, r \geqslant 1$ *and let* $M$ *be a finite dimensional* $\mathrm{GL}_{n(r)}$-*module. Then* $\mathrm{H}^*(\mathrm{GL}_{n(r)}, k)$ *is a finitely generated algebra and* $\mathrm{H}^*(\mathrm{GL}_{n(r)}, M)$ *is a finite module over* $\mathrm{H}^*(\mathrm{GL}_{n(r)}, k)$.

*Sketch of proof*. — As in the proof of Theorem 4.1, we analyze the May spectral sequences of Theorem 2.10, $\{E_r^{*,*}(k)\}$ for $\mathrm{H}^*(\mathrm{GL}_{n(r)}, k)$ and $\{E_r^{*,*}(M)\}$ for $\mathrm{H}^*(\mathrm{GL}_{n(r)}, M)$.

Let

$$S^*(\mathrm{gl}_n^{(r)\#}[2p^{r-i}]) \subset S^*(\mathrm{gl}_n^{(i)\#}[2])$$

denote the poynomial subalgebra generated by the subspace $\mathrm{gl}_n^{(r)} \subset S^{p^{r-i}}(\mathrm{gl}_n^{(i)\#}[2])$ of $p^{r-i}$-th powers of $\mathrm{gl}_n^{(i)\#}[2]$. Then Corollary 4.5 together with the evident naturality of our constructions with respect to $\mathrm{GL}_{n(r)} \to \mathrm{GL}_{n(r)}/\mathrm{GL}_{n(r-1)} \simeq \mathrm{GL}_{n(1)}$ implies that

$$\bigotimes_{i=1}^{r} S^*(\mathrm{gl}_n^{(r)\#}[2p^{r-i}]) \otimes M \subset E_1^{*,*}(M)$$

consists of permanent cycles.

Clearly, $E_1^{*,*}(M)$ is a finite $\otimes_{i=1}^r S^*(\mathrm{gl}_n^{(r)\#}[2p^{r-i}])$ module. As argued in the proof of Theorem 4.1, this implies that $E_\infty^{*,*}(k)$ is a finitely generated algebra and thus also $\mathrm{H}^*(\mathrm{GL}_{n(r)}, k)$ is also finitely generated. Since the action of $\otimes_{i=1}^r S^*(\mathrm{gl}_n^{(r)\#}[2p^{r-i}])$ on $\mathrm{H}^*(\mathrm{GL}_{n(r)}, M)$ factors through $\mathrm{H}^*(\mathrm{GL}_{n(r)}, k)$, we conclude that $\mathrm{H}^*(\mathrm{GL}_{n(r)}, M)$ is a finite $\mathrm{H}^*(\mathrm{GL}_{n(r)}, k)$ module.                                          $\square$

We are now in a position to outline the remainder of the proof of finite generation of $\mathrm{H}^*(G, k)$ for an arbitrary finite group scheme. This proof relies on earlier work of L. Evens [6] who, together with B. Venkov [19], proved the finite generation of the cohomology algebra of a finite group.

***Theorem 4.7*** ([**12**, 1.1]). — *Let $G$ be a finite group scheme and let $M$ be a finite dimensional $G$-module. Then $\mathrm{H}^*(G,k)$ is a finitely generated algebra and $\mathrm{H}^*(G,M)$ is a finite module over $\mathrm{H}^*(G,k)$.*

*Outline of proof.* — If $G$ an an infinitesimal group scheme of height $\leqslant r$, then $G$ admits an embedding as a closed subgroup scheme of $\mathrm{GL}_{n(r)}$. Shapiro's Lemma,

$$\mathrm{H}^*(G,M) \simeq \mathrm{H}^*(\mathrm{GL}_{n(r)}, \mathrm{Ind}_G^{\mathrm{GL}_{n(r)}} M),$$

in conjunction with Theorem 4.6 easily implies the assertions of the theorem for such infinitesimal group schemes $G$.

For applications considered in the next lecture, we utilize a different proof of finite generation for $G$ infinitesimal. Namely, a closed embedding $G \subset \mathrm{GL}_{n(r)}$ induces a map of spectral sequences

$$E_1^{*,*}(\mathrm{GL}_{n(r)},k) = \bigotimes_1^r S^*(\mathrm{gl}_n^{(i)\#}[2]) \otimes \bigotimes_1^r \Lambda^*(\mathrm{gl}_n^{(i-1)\#}[1])$$

$$\downarrow$$

$$E_1^{*,*}(G,k) = \bigotimes_1^r S^*(g^{(i)\#}[2]) \otimes \bigotimes_1^r \Lambda^*(g^{(i-1)\#}[1])$$

which is surjective on $E_1^{*,*}$. Thus, the argument given in the proof of Theorem 4.6 applies, since it suffices to show that $\mathrm{H}^*(G,k)$ is a finite module over $\otimes_{i=1}^r S^*(\mathrm{gl}_n^{(r)\#}[2p^{r-i}])$ and is thus finitely generated. Similarly, $\mathrm{H}^*(G,M)$ is finite as a $\otimes_{i=1}^r S^*(\mathrm{gl}_n^{(r)\#}[2p^{r-i}])$ module and thus also as a $\mathrm{H}^*(G,k)$ module.

Since $\mathrm{H}^*(G,M) \otimes_k K = \mathrm{H}^*(G_K, M \otimes_k K)$ for any field extension $K/k$, to prove finite generation for an arbitrary finite group scheme we may assume that $k$ is algebraically closed. In this case, the split extension

$$1 \longrightarrow G^o \longrightarrow G \longrightarrow \pi_0(G) \longrightarrow 1$$

is necessarily a semi-direct product of an infinitesimal group scheme and a finite group. Then, one readily adapts results of [**6**] to conclude finite generation for $G$ knowing finite generation for $G^o$ and using the Evens-Venkov theorem asserting finite generation for $\pi_0(G)$. (see [**12**, 1.9,1.10] for details). □

## 5. Qualitative description of $\mathrm{H}^{\mathrm{ev}}(G,k)$

Because $\mathrm{H}^*(G,k)$ is graded commutative, the subalgebra $\mathrm{H}^{\mathrm{ev}}(G,k)$ of even dimensional elements is a commutative algebra. Indeed, for $p = 2$, $\mathrm{H}^*(G,k)$ is already commutative, but for simplicity we ignore this finer commutative algebra and consider $\mathrm{H}^{\mathrm{ev}}(G,k)$ for any prime $p \geqslant 2$. In [**15**], D. Quillen described the maximal ideal spectrum $|\pi|$ of the commutative algebra $\mathrm{H}^{\mathrm{ev}}(\pi,k)$ for a finite group $\pi$ in terms of the elementary abelian $p$-subgroups of $\pi$. This remains a remarkable work, both for introducing the possibility of identifying the maximal ideal spectrum as well as for the completeness of the result. For example, Quillen gives us an explicit description of the maximal ideal spectrum of $\mathrm{H}^{\mathrm{ev}}(\mathrm{GL}_n(\mathbb{F}_q),k)$, $q=p^d$, even though we know very little about the individual cohomology groups $\mathrm{H}^i(\mathrm{GL}_n(\mathbb{F}_q),k)$. (For example, we do not even know what is the smallest positive degree such that $\mathrm{H}^i(\mathrm{GL}_n(\mathbb{F}_q),k)\neq 0$.) It is interesting to note that Quillen also observed that $\mathrm{H}^i(\mathrm{GL}_\infty(\mathbb{F}_q),k)=0$ for $i>0$, a fact which is closely related to the fact that $k=S^0\in\mathcal{P}$ is acyclic.

**Theorem 5.1 ([15]).** — *Let $\pi$ be a finite group, assume that $k$ is algebraically closed, and let $|\pi|$ denote the maximal ideal spectrum of $\mathrm{H}^{\mathrm{ev}}(\pi,k)$. Then the natural map*

$$\varinjlim_{\{E\to\pi\}}|E|\longrightarrow|\pi|$$

*is a homeomorphism, where the indexing category for the colimit is the category whose objects are elementary abelian subgroups of $\pi$ and whose maps are compositions of group inclusions and maps induced by conjugations by elements of $\pi$.*

Recall that if $E$ is an elementary abelian $p$-group of rank $n$, then $\mathrm{H}^*(E,k)\simeq k[x_1,\ldots,x_n]\otimes\Lambda(y_1,\ldots,y_n)$ where each $x_i\in\mathrm{H}^2(E,k),y_i\in\mathrm{H}^1(E,k)$ for $p\neq 2$ (for $p=2$, $\mathrm{H}^*(E,k)\simeq k[y_1,\ldots,y_n]$ with each $y_i\in\mathrm{H}^1(E,k)$). Thus, $|E|$ is an affine space of dimension $n$. Theorem 5.1 tells us that the Krull dimension of the commutative ring $\mathrm{H}^{\mathrm{ev}}(\pi,k)$ equals the maximal rank among elementary abelian $p$-subgroups of $\pi$. We can restate Theorem 5.1 as asserting that $|\pi|$ is the identification space of the following projection

$$(5.2)\qquad\coprod_{E\ max}|E|/W_E\longrightarrow|\pi|,$$

where the coproduct is indexed by conjugacy classes of maximal elementary abelian $p$-subgroups of $\pi$ and where $W_E$ denotes the normalizer of $E$ modulo its centralizer as a subgroup of $\pi$. Moreover, points of $e\in|E|/W_E$, $e'\in|E'|/W_{E'}$ are mapped via (5.2) to the same point of $|\pi|$ if and only if there exist conjugates $\tilde E,\tilde E'$ of $E,E'$ and a point $e''\in|\tilde E\cap\tilde E'|/W_{\tilde E\cap\tilde E'}$ mapping to $e,e'$.

To prove Theorem 5.1, Quillen proves i) that the map from the coproduct is surjective by showing that any cohomology class $\zeta\in\mathrm{H}^*(\pi,k)$ which restricts to $0\in\mathrm{H}^*(E,k)$ for every elementary abelian subgroup $E\subset\pi$ is nilpotent; ii) that any point of $|E|/W_E$ not in the image of $|E'|/W_{E'}$ with $E'$ a proper subgroup of $E$ maps injectively into $|\pi|$ by showing that any class in a certain localization of $\mathrm{H}^{\mathrm{ev}}(E,k)^{W_E}$ admits a $p$-th power in the image of $\mathrm{H}^{\mathrm{ev}}(\pi,k)$.

As first observed by Friedlander-Parshall, the maximal ideal spectrum of the even dimensional cohomology of a finite dimensional restricted Lie algebra also has an explicit description. Conditions on the prime $p$ required by Friedlander-Parshall were relaxed by Andersen-Janzten and eliminated altogether by Suslin-Friedlander-Bendel.

**Theorem 5.3 (cf. [10], [1], [18]).** — *Let $G$ be an infinitesimal group scheme of height 1, let $g=\mathrm{Lie}\,G$, and assume that $k$ is algebraically closed. Denote by $N_p(g)\subset g$ the* *p-nilpotent cone of g, the set of elements $x \in g$ satisfying $x^{[p]} = 0$. Then there is a natural homeomorphism*

$$\Psi : N_p(G) \overset{\sim}{\longrightarrow} |G|$$

*where $|G|$ denotes the maximal ideal spectrum of $\mathrm{H}^{ev}(G, k)$.*

Theorem 5.3 was generalized to arbitrary infinitesimal group schemes in two papers by Suslin-Friedlander-Bendel [**17**], [**18**] in a form which is more precise even in the height 1 case. (Namely, these papers deal with schemes rather than maximal ideal spectra. Among other advantages, this permits them to consider an arbitrary field $k$.) The schemes that generalize the variety $N_p(g)$ of Theorem 5.3 are introduced in the next proposition.

***Proposition 5.4*** ([**17**, 1.5]). — *Let $G$ be an affine group scheme. Then the functor on commutative $k$-algebras*

$$A \longmapsto \mathrm{Hom}_{Grps/A}(\mathbb{G}_{a(r)} \otimes A \longrightarrow G \otimes A)$$

*is representable by an affine scheme $V_r(G)$.*
*For $G = \mathrm{GL}_n$,*

$$V_r(\mathrm{GL}_n)(k) = \{(\alpha_1, \ldots, \alpha_r) \in M_n(k)^r \big| \alpha_i^p = 0 = [\alpha_i, \alpha_j]\}.$$

*In the case $r = 1$, $V_1(G)$ is the scheme whose underlying variety is the p-nilpotent cone $N_p(\mathrm{Lie}(G))$ considered in Theorem 5.3.*

We call a homomorphism $\alpha : \mathbb{G}_{a(r)} \otimes A \to G \times A$ a *1-parameter subgroup* of height $r$ defined over $A$.

Recall from Example 1.5 that $k\mathbb{G}_{a(r)} = k[X_1, \ldots, X_r]/(X_i^p)$ where $X_i$ is the operator $\dfrac{1}{p^{i-1}!}\dfrac{d^{p^{i-1}}}{dt^{p^{i-1}}}$ on $k[\mathbb{G}_{a(r)}] = k[t]/t^{p^r}$. We consider the map

$$\varepsilon : k\mathbb{G}_{a(1)} \longrightarrow \mathbb{G}_{a(r)}, \quad u \longmapsto X_r,$$

where $u \in k\mathbb{G}_{a(1)}$ is the dual of $t \in k[t]/t^p = k[\mathbb{G}_{a(1)}]$. So defined, $\varepsilon$ is not a map of Hopf algebras (*i.e.*, does not commute with the coproduct), but does induce a map on cohomology

$$\varepsilon^* : \mathrm{H}^*(\mathbb{G}_{a(r)}, k) \longrightarrow \mathrm{H}^*(\mathbb{G}_{a(1)}, k).$$

Observe that a 1-parameter subgroup $\alpha : \mathbb{G}_{a(r)} \to G$ determines a homomorphism of graded algebras

$$\varepsilon^* \circ \alpha^* : \mathrm{H}^{ev}(G, k) \longrightarrow \mathrm{H}^{ev}(\mathbb{G}_{a(r)}, k) \longrightarrow \mathrm{H}^{ev}(\mathbb{G}_{a(1)}, k) \simeq k[t].$$

This determines a natural set-theoretic map

$$V_r(G) \longrightarrow \mathrm{Spec}\, \mathrm{H}^{ev}(G, k), \quad \alpha \longmapsto \ker\{\varepsilon^* \circ \alpha^*\},$$

where $\mathrm{Spec}\, \mathrm{H}^{ev}(G, k)$ denotes the prime ideal spectrum of the graded algebra $\mathrm{H}^{ev}(G, k)$. The following proposition asserts that this set-theoretic map admits a natural refinement as a map of schemes.

**Proposition 5.5** ([**17**, 1.14]). — *For any affine group scheme $G$, there is a natural homomorphism of graded commutative $k$-algebras*

$$(5.6) \qquad\qquad \psi : \mathrm{H}^{\mathrm{ev}}(G,k) \longrightarrow k[V_r(G)]$$

*which multiplies degrees by $p^r/2$.*

*In the case $r = 1$ and $k$ algebraically closed, the map on affine varieties induced by $\psi$ is the homeomorphism $\Psi$ of Theorem 5.3.*

*Proof.* — The construction of this map is of sufficient independent interest that we sketch it here. Let

$$u : \mathbb{G}_{a(r)} \otimes k[V_r(G)] \longrightarrow G \otimes k[V_r(G)]$$

correspond to

$$\mathrm{Id}_{k[V_r(G)]} \in V_r(G)(k[V_r(G)]) = \mathrm{Hom}_{Grps/k[V_r(G)]}(\mathbb{G}_{a(r)} \otimes k[V_r(G)], G \otimes k[V_r(G)]).$$

Consider

$$u^* : \mathrm{H}^*(G,k) \longrightarrow \mathrm{H}^*(G,k) \otimes k[V_r(G)] = \mathrm{H}^*(G \otimes k[V_r(G)], k[V_r(G)])$$
$$\longrightarrow \mathrm{H}^*(\mathbb{G}_{a(r)} \otimes k[V_r(G)], k[V_r(G)]) = \mathrm{H}^*(\mathbb{G}_{a(r)}, k) \otimes k[V_r(G)].$$

For any $\zeta \in \mathrm{H}^{2j}(G,k)$, we define $\psi(\zeta)$ to be the coefficient of $x_r^j$, where $x_r = \varepsilon^*(x) \in \mathrm{H}^2(\mathbb{G}_{a(r)},k)$ with $x \in \mathrm{H}^2(\mathbb{G}_{a(1)},k)$ the chosen polynomial generator. $\qquad\square$

We proceed to outline how Suslin-Friedlander-Bendel construct an "inverse modulo $p$-nilpotents" of $\psi$, thereby verifying that $\psi$ determines a homeomorphism $\Psi : V_r(G) \to \mathrm{Spec}\, \mathrm{H}^{\mathrm{ev}}(G,k)$ of prime ideal spectra. The following theorem explicitly exhibits such an "inverse" for $\psi$ in the special case of $G = \mathrm{GL}_{n(r)}$.

**Theorem 5.7** ([**17**, 5.2]). — *The fundamental classes $e_i \in \mathrm{H}^{2p^{i-1}}(\mathrm{GL}_n, \mathfrak{gl}_n^{(i)})$ determine a map of algebras*

$$\otimes_{i=1}^r e_i^{(r-i)} : \bigotimes_{i=1}^r S^*(\mathfrak{gl}_n^{(r)\#}[2p^{i-1}]) \to \mathrm{H}^{\mathrm{ev}}(\mathrm{GL}_{n(r)},k)$$

*which factors through the quotient map $\otimes_{i=1}^r S^*(\mathfrak{gl}_n^{(r)\#}[2p^{i-1}]) \to k[V_r(G)]$ associated to the embedding of $V_r(\mathrm{GL}_n) \subset (M_n)^r$. Thus, $\otimes_{i=1}^r e_i^{(r-i)}$ determines a map*

$$\phi : k[V_r(\mathrm{GL}_n)] \longrightarrow \mathrm{H}^{\mathrm{ev}}(\mathrm{GL}_{n(r)},k).$$

*Moreover, the composition*

$$\psi \circ \phi : k[V_r(\mathrm{GL}_n)] \longrightarrow \mathrm{H}^*(\mathrm{GL}_{n(r)},k) \longrightarrow k[V_r(\mathrm{GL}_n)],$$

*equals $F^r$, the $r$-th iterate of the (geometric) Frobenius sending generators of the $k$-algebra $k[V_r(\mathrm{GL}_n)]$ to their $p^r$-th power.*

As in the proof of finite generation, establishing a qualitative description for $H^{ev}(GL_{n(r)}, k)$ goes a long way toward establishing a similar description of infinitesimal groups of height $\leqslant r$. In particular, Theorem 5.7 together with the naturality of the May spectral sequence (see (2.10)) easily implies the following corollary.

**Corollary 5.8**. — *For any infinitesimal group scheme $G$ of height $\leqslant r$, $\psi : H^{ev}(G, k) \to k[V_r(G)]$ has image containing $F^r(k[V_r(G)]) \subset k[V_r(G)]$. In particular, $\psi$ is surjective modulo p-th powers.*

To complete the qualitative description of $H^{ev}(G, k)$ for $G$ infinitesimal we must show that $\psi$ is "injective modulo nilpotents". This is verified by showing that a class $\zeta \in H^{ev}(G, k)$ which restricts to 0 via every 1-parameter subgroup is nilpotent, a result analogous to Quillen's result asserting the cohomology of $H^*(\pi, k)$ is detected modulo nilpotents by restrictions to elementary abelian subgroups of a finite group $\pi$.

Thus, Suslin-Friedlander-Bendel conclude the following.

**Theorem 5.9** (**[18**, 5.2]). — *Let $G$ be an infinitesimal group of height $\leqslant r$. Then the map of affine schemes associated to (5.6),*

$$\Psi : V_r(G) \longrightarrow \text{Spec}\, H^{ev}(G, k),$$

*is a homeomorphism.*

Quite recently, Friedlander and J. Pevtsova have introduced a qualitative description of $H^*(G, k)$ for any finite group scheme which encompasses the case of finite groups presented in Theorem 5.1 and that of infinitesimal group schemes presented in Theorem 5.9. This generalization loses the scheme-theoretic information of Theorem 5.9 and requires the assumption that $k$ be algebraically closed.

**Definition 5.10**. — Let $G$ be a finite group scheme over the algebraically closed field $k$. An abelian $p$-point of $G$ is a flat map of algebras $\alpha : k[u]/u^p \to kG$ which factors through some abelian subgroup scheme of $G$. Two such abelian $p$-points $\alpha, \beta$ are said to be equivalent provided that they satisfy the following condition: for every finite dimensional $G$-module $M$, $\alpha^*(M)$ is projective (as a $k[u]/u^p$-module) if and only $\beta^*(M)$ is projective.

The set of equivalence classes of abelian $p$-points of $G$ is denoted $P(G)$. This set is given a topology by defining a subset $Y \subset P(G)$ to be closed if and only if there exists a finite dimensional module $M$ such that $Y$ consists of those equivalence classes of abelian $p$-points $\alpha$ for which $\alpha^*(M)$ is not projective.

The primary motivation for the above definition is the consideration of "support varieties" for $G$-modules, a topic which we do not consider for lack of time but which is a natural extension to the subject matter of this lecture. However, Definition 5.10 does enable us to formulate the following qualitative description of $H^{ev}(G, k)$ for an arbitrary finite group scheme $G$.

**Theorem 5.11** ([**11**, 4.8]). — *Let $G$ be a finite group scheme over an algebraically closed field $k$. Then there is a natural homeomorphism*

$$P(G) \xrightarrow{\sim} \mathrm{Proj}(|G|),$$

*sending an abelian p-point $\alpha : k[u]/u^p \to \mathrm{H}^{\mathrm{ev}}(G, k)$ to the homogeneous ideal $\ker\{\alpha^*\}$.*

To prove Theorem 5.11, Friedlander and Pevtsova use the following theorem recently proved by A. Suslin extending a result by C. Bendel [**2**] which itself extended results of Suslin-Friedlander-Bendel. We say that a finite group scheme is *quasi-elementary* if it is isomorphic to a product of the form $\mathbb{G}_{a(r)} \times \mathbb{Z}/p^s$ for some $r, s \geqslant 0$.

**Theorem 5.12 (A. Suslin [16])**. — *Let $G$ be a finite group scheme, $\Lambda$ be a unital associative $G$-algebra, and $\zeta \in \mathrm{H}^{\mathrm{ev}}(G, \Lambda)$ be a homogeneous cohomology class of even degree. Then $\zeta$ is nilpotent if and only if $\zeta_K$ restricts to a nilpotent class in $\mathrm{H}^{\mathrm{ev}}(\mathcal{E}_K, \Lambda_K)$ for every field extension $K/k$ and every quasi-elementary subgroup scheme $\mathcal{E}_K$ of $G_K$.*

*Acknowledgements*. — We thank the organizers of the Nantes conference for the opportunity to visit Nantes and participate in that very successful conference. We also take this opportunity to thank Andrei Suslin for sharing with us many fundamental ideas he has contributed to the study of the cohomology of finite group schemes. Finally, we are especially grateful to Julia Pevtsova who corrected many errors in a preliminary draft of these notes.

# References

[1] H. ANDERSEN & J. JANTZEN – Cohomology of induced representations for algebraic groups, *Math. Ann.* **269** (1985), p. 487–525.

[2] C. BENDEL – Cohomology and Projectivity, *Math. Proc. Cambridge Philos. Soc.* **131** (2001), p. 405–425.

[3] P. CARTIER – Une nouvelle opération sur les formes différentielles, *C. R. Acad. Sci. Paris Sér. I Math.* **244** (1957), p. 426–428.

[4] E. CLINE, B. PARSHALL, L. SCOTT & W. VAN DER KALLEN – Rational and generic cohomology, *Invent. Math.* **39** (1977), p. 143–163.

[5] S. DONKIN – On Schur algebras and related algebras, I, *J. Algebra* **104** (1986), p. 310–328.

[6] L. EVENS – The cohomology ring of a finite group, *Trans. Amer. Math. Soc.* **101** (1961), p. 224–239.

[7] V. FRANJOU, E. FRIEDLANDER, A. SCORICHENKO & A. SUSLIN – General linear and functor cohomology over finite fields, *Ann. of Math. (2)* **150** (1999), p. 663–728.

[8] V. FRANJOU, J. LANNES & L. SCHWARTZ – Autour de la cohomologie de MacLane des corps finis, *Invent. Math.* **115** (1994), p. 513–538.

[9] E. FRIEDLANDER & B. PARSHALL – On the cohomology of algebraic and related finite groups, *Invent. Math.* **74** (1983), p. 85–117.

[10] _____, Cohomology of Lie algebras and Algebraic groups, *Amer. J. Math.* **108** (1986), p. 235–253.

[11] E. FRIEDLANDER & J. PEVTSOVA – Representation-theoretic support spaces for finite group schemes, submitted for publication.

[12] E. FRIEDLANDER & A. SUSLIN – Cohomology of finite group schemes over a field, *Invent. Math.* **127** (1997), p. 209–270.

[13] J.C. JANTZEN – *Representations of Algebraic groups*, Academic Press, 1987.

[14] T. PIRASHVILI – Higher additivizations, *Trudy Tbiliss. Mat. Inst. Razmadze Akad. Nauk Gruzin. SSR* **91** (1988), p. 44–54, Russian.

[15] D. QUILLEN – The spectrum of an equivariant cohomology ring: I, II, *Ann. of Math.* **94** (1971), p. 549–572, 573–602.

[16] A. SUSLIN – The detection theorem for finite group schemes, in preparation.

[17] A. SUSLIN, E. FRIEDLANDER & C. BENDEL – Infinitesimal 1-parameter subgroups and cohomology, *J. Amer. Math. Soc.* **10** (1997), p. 693–728.

[18] ———, Support varieties for infinitesimal group schemes, *J. Amer. Math. Soc.* **10** (1997), p. 729–759.

[19] B. VENKOV – Cohomology algebras for some classifying spaces, *Dokl. Akad. Nauk SSSR* **127** (1959), p. 943–944.

*Panoramas & Synthèses*
**16**, 2003, p. 55–100

# ALGÈBRE DE STEENROD, MODULES INSTABLES
# ET FONCTEURS POLYNOMIAUX

*par*

Lionel Schwartz

**Résumé.** — Ce texte donne une introduction aux propriétés algébriques de l'algèbre de Steenrod et de la catégorie des modules instables. Les relations avec la catégorie des foncteurs polynomiaux sont établies.

**Abstract (The Steenrod algebra, unstable modules and polynomial functors).** — This text gives an introduction to the algebraic properties of the Steenrod algebra and of the category of unstable modules. The link with the category of polynomial functors is described.

## 0. Introduction

Ces notes ont leur origine dans la session *État de la Recherche* autour des foncteurs polynomiaux et des modules instables. À la différence des conférences de l'auteur, elles présentent l'algèbre de Steenrod *via* le schéma en groupes des automorphismes du groupe formel additif. Cette présentation ne fait pas appel à la topologie et n'utilise que de l'algèbre élémentaire. L'étude algébrique des modules instables sur l'algèbre de Steenrod qui suit permet de montrer leur similitude avec les foncteurs polynomiaux.

Si ce texte est surtout un exposé actuel de la théorie, il contient aussi quelques nouveautés, en particulier des généralisations au cas $p > 2$ de résultats connus pour $p = 2$ (qui, si elles appartiennent au folklore, ne sont pas écrites dans des sources accessibles) :

– l'extension de l'approche évoquée plus haut *via* le groupe formel additif à $p > 2$ ;
– un énoncé, pour $p > 2$, correspondant à l'identification du module $F(n)$ aux invariants sous le groupe symétrique dans $F(1)^{\otimes n}$ pour $p = 2$ ;
– l'introduction des modules instables libres ;
– un nouveau calcul de l'algèbre de Miller $J_*^*$.

*Classification mathématique par sujets* **(2000).** — 55-02, 55S10.
*Mots clefs.* — Algèbre de Steenrod, modules instables, foncteurs polynomiaux.

**Notations.** — Les notations suivantes sont utilisées dans la suite comme dans les autres textes de ce volume. La catégorie des $k$-espaces vectoriels est notée **Vect**, celle des espaces vectoriels de dimension finie $\mathcal{V}_k^f$ ou $\mathcal{V}^f$. Si $V$ est un espace vectoriel sur le corps $k$, $S^n(V)$ est la $n$-ième puissance symétrique, $\Lambda^n(V)$ est la $n$-ième puissance extérieure, $\Gamma^n(V)$ est la $n$-ième puissance divisée. Le groupe symétrique est noté $\mathfrak{S}_n$. On rappelle que $\Gamma^k(V)$ est isomorphe à $(V^{\otimes k})^{\mathfrak{S}_k}$ (les invariants sous l'action du groupe symétrique) et que, $V$ étant de dimension finie, on a un isomorphisme naturel :

$$S^n(V) \cong \Gamma^n(V^{\#})^{\#}.$$

L'application $\mathbf{V}$, dite Verschiebung, qui envoie $\Gamma^{np}(V)$ dans $\Gamma^n(V)$, est duale du morphisme de Frobenius : $x \mapsto x^p$ de $S^*(V^{\#})$ dans lui même. Pour $p = 2$, elle est caractérisée de la manière suivante : si $\delta$ désigne la diagonale de l'algèbre de Hopf $\Gamma^*(V)$, si l'on écrit :

$$\delta(x) = \sum_i (a_i \otimes b_i + b_i \otimes a_i) + z \otimes z,$$

l'élément $z$ est bien déterminé, et $\mathbf{V}(x) = z$.

Dans tout ce qui suit on se place sur un corps premier $\mathbb{F}_p$ ; cependant, d'après [**12**], les résultats s'étendent sans difficulté à un corps fini quelconque.

## 1. Le groupe additif et le dual de Milnor

**1.1. Le groupe additif.** — Une manière (due à J. Morava et P. Cartier) d'introduire l'algèbre de Steenrod est de commencer par décrire son dual. Celui-ci apparaît naturellement en considérant les automorphismes du groupe formel additif $\mathbb{G}_a$, plus précisément de la complétion formelle $\widehat{\mathbb{G}}_a$ de $\mathbb{G}_a$. Une *loi de groupe formel* commutatif, à coefficients dans un anneau $A$, est une série formelle $F(X, Y) \in A[[X, Y]]$ qui satisfait aux propriétés suivantes :

- $F(X, 0) = F(0, X) = X$,
- $F(X, Y) = F(Y, X)$,
- $F(X, F(Y, Z)) = F(F(X, Y), Z)$.

Rappelons aussi qu'une $A$-algèbre de Hopf est la donnée :

- d'une $A$-algèbre $L$, d'unité $\eta : A \to L$ et de multiplication $\mu$ ;
- d'une application d'algèbres $\delta : L \to L \otimes L$ qui admet une coünité $\varepsilon : L \to A$, et est coassociative.

Les secondes conditions signifient que :

- $\varepsilon \circ \eta$ est l'identité de $A$ ;
- que : $\mu \circ (\varepsilon \otimes \mathrm{Id}_L) \circ \delta = \mathrm{Id}_L$ et $\mu \circ (\mathrm{Id}_L \otimes \varepsilon) \circ \delta = \mathrm{Id}_L$ ;
- enfin que : $(\delta \otimes \mathrm{Id}_L) \circ \delta = (\mathrm{Id}_L \otimes \delta) \circ \delta$.

Si, de plus, $\delta = \tau \circ \delta$, où $\tau$ est l'échange des deux facteurs du produit tensoriel, l'algèbre est dite cocommutative.

L'anneau $A[[X]]$ est muni de la topologie $X$-adique, et l'anneau $A[[X,Y]]$ de la topologie $(X,Y)$-adique. Les applications d'algèbres considérées ci-dessous sont toujours supposées continues pour ces topologies. La définition d'une algèbre de Hopf (donnée ci-dessus) est ainsi étendue à ce contexte topologique. Les relations définissant un groupe formel $F$ expriment que $A[[X]]$, muni de la diagonale :

$$\delta : A[[X]] \longrightarrow A[[X_1, X_2]]^\wedge$$

définie par : $\delta(X) = F(X_1, X_2)$, est une algèbre de Hopf topologique cocommutative. Ci-dessus, on a dû compléter l'anneau $A[[X_1, X_2]]$ de manière *ad hoc*. L'existence de la coünité résulte de la première condition, la cocommutativé résulte de la seconde, et la coassociativité de la troisième.

Un endomorphisme du groupe formel est, par définition, un endomorphisme continu de la $A$-algèbre de Hopf $A[[X]]$. Une telle application est entièrement déterminée par l'image de $X$, qui est une série formelle $\ell(X)$, sans terme constant, telle que :

$$F(\ell(X), \ell(Y)) = \ell(F(X,Y)).$$

On vérifie facilement que si à $\phi$ et $\psi$ sont associées respectivement les séries formelles $h$ et $\ell$, à $\psi \circ \phi$ correspond $h \circ \ell$. Dans le cas du groupe additif $\mathbb{G}_a(X,Y) = X + Y$ (ou de sa complétion formelle) la série formelle associée à un endomorphisme est donc telle que : $\ell(X+Y) = \ell(X) + \ell(Y)$. Si $L$ est une $\mathbb{F}_2$-algèbre, les séries formelles qui vérifient cette équation sont de la forme :

$$\sum_i a_i X^{2^i}, \quad a_i \in L.$$

Un tel endomorphisme est un automorphisme si, et seulement si, $a_0$ est une unité de $L$. Un automorphisme est dit *spécial* si $a_0 = 1$. Le groupe des automorphismes spéciaux est noté simplement $\Gamma(L)$ pour $\widetilde{\mathrm{Aut}}_{\mathbb{G}_a}(L)$. On a une équivalence naturelle

$$\Gamma(L) \cong \mathrm{Hom}_{\mathbb{F}_2\text{-Alg}}(\mathbb{F}_2[\xi_i], L),$$

qui, à $\phi \in \mathrm{Hom}_{\mathbb{F}_2\text{-Alg}}(\mathbb{F}_2[\xi_i], L)$, associe $\ell_\phi = X + \sum_{i>0} \phi(\xi_i) X^{2^i}$. On note $\mathcal{A}_2^*$ l'algèbre $\mathbb{F}_2[\xi_i]$, l'algèbre des fonctions régulières sur le schéma en groupes $\Gamma$.

Le fait que le foncteur $\Gamma$ prenne valeurs dans la catégorie des groupes implique que l'algèbre $\mathcal{A}_2^* = \mathbb{F}_2[\xi_i]$ hérite d'une structure d'algèbre de Hopf. La diagonale $\delta$ est le produit $g * d$, pour la structure de groupe, des applications canoniques $g, d :$ $\mathcal{A}_2^* \to (\mathcal{A}_2^*)^{\otimes 2}$ données respectivement par $x \mapsto x \otimes 1$ et $x \mapsto 1 \otimes x$. Les propriétés de coassociativité et coünité de $\delta$ se déduisent directement des propriétés d'associativité et d'unité du groupe $\Gamma((\mathcal{A}_2^*)^{\otimes 2})$.

**Théorème 1.1.** — *On a :*
$$\delta(\xi_\ell) = \sum_{\substack{i+j=\ell, \\ i,j \geqslant 0}} \xi_i^{2^j} \otimes \xi_j.$$

*Dans cette formule, par convention, $\xi_0 = 1$.*

*Démonstration*. — Si $\phi$ et $\psi$ sont des endomorphismes, on a : $\ell_{\phi*\psi} = \ell_\psi \circ \ell_\phi$. Or, par définition :

$$\ell_g = X + \sum_{i>0} (\xi_i \otimes 1) X^{2^i},$$

et

$$\ell_d = X + \sum_{i>0} (1 \otimes \xi_i) X^{2^i}.$$

Il en résulte, en posant $\xi_0 = 1$ :

$$\ell_\delta = \ell_{g*d} = \ell_d \circ \ell_g = \sum_{j\geqslant 0} 1 \otimes \xi_j \Big( \sum_{i\geqslant 0} (\xi_i \otimes 1) X^{2^i} \Big)^{2^j} = \sum_{i,j\geqslant 0} (\xi_i^{2^j} \otimes \xi_j) X^{2^{i+j}}. \qquad \square$$

Introduisons une graduation sur $\mathcal{A}_2^*$ en posant $|\xi_i| = -2^i + 1$. L'application $\delta$ respecte cette graduation. Considérons donc $\mathcal{A}_2^*$ comme algèbre de Hopf graduée.

***Définition 1.2***. — L'algèbre de Steenrod modulo 2 $\mathcal{A}_2$ est l'algèbre de Hopf graduée duale de $\mathcal{A}_2^*$.

Précisons que, par définition, $\mathcal{A}_2^k = ((\mathcal{A}_2^*)_{-k})^\#$. Ainsi, l'algèbre graduée $\mathcal{A}_2$ est non nulle en degrés positifs ou nuls. Indexons la base monomiale de l'algèbre $\mathcal{A}_2^*$ par les multi-indices, en notant $\xi^R$ le monôme $\xi_1^{r_1}, \ldots \xi_h^{r_h}$ pour un multi-indice $R = (r_1, \ldots, r_h)$. Une base graduée de $\mathcal{A}_2$ est obtenue par dualisation :

***Définition 1.3***. — Soit $\{\xi^R \mid h \geqslant 0, R = (r_1, \ldots, r_h), r_i \geqslant 0, r_h > 0\}$ la base monomiale de $\mathcal{A}_2^*$. Les éléments de sa base duale sont notés $\mathrm{Sq}(R)$. Cette base est appelée base de Milnor de $\mathcal{A}_2$. Les éléments $\mathrm{Sq}(i)$ sont notés $\mathrm{Sq}^i$.

***Définition 1.4***. — L'excès de l'opération $\mathrm{Sq}(r_1, \ldots, r_h)$ est l'entier $r_1 + \cdots + r_h$.

***Remarque 1.5***. — L'homomorphisme de $\mathbb{F}_2$-algèbres de $M$ vers $\mathbb{F}_2$, qui envoie $\xi_1$ vers 1 et les autres générateurs $\xi_i$, $i > 1$, vers 0, n'est pas un élément de ce dual, mais du produit $\prod_i \mathcal{A}_2^i$ : c'est « le produit » des formes linéaires prenant la valeur 1 sur $\xi_1^i$ et la valeur 0 sur les autres mônomes. Cette opération est appelée opération de Steenrod totale. On la note dans ce texte $\mathbf{Sq}^{(1)} = 1 + \mathrm{Sq}^1 + \mathrm{Sq}^2 + \cdots + \mathrm{Sq}^i + \cdots$, ou simplement $\mathbf{Sq}$.

On peut définir plus généralement $\mathbf{Sq}^{(j)}$ comme étant l'homomorphisme de $\mathbb{F}_2$-algèbres de $\mathcal{A}_2^*$ vers $\mathbb{F}_2$ qui envoie $\xi_j$ vers 1 et les autres générateurs $\xi_i$, $i \neq j$, vers 0.

**1.2. Le cas $p > 2$.** — Tout ce qu'on vient de dire s'étend à la caractéristique $p > 2$. Soit $\Gamma_p(L)$ le schéma en groupes des séries formelles de la forme $X + a_1 X^p + \cdots + a_k X^{p^k} + \cdots$, où les $a_i$ appartiennent à une $\mathbb{F}_p$-algèbre. On a :

***Théorème 1.6***. — *L'algèbre des fonctions régulières sur le schéma en groupes $L \mapsto \Gamma_p(L) = \widetilde{\mathrm{Aut}}_{\mathbb{G}_a}(L)$ est isomorphe à $\mathbb{F}_p[\xi_i, i \geqslant 0]$. Sa structure d'algèbre de Hopf est* *déterminée par :*

$$\delta(\xi_\ell) = \sum_{\substack{i+j=\ell, \\ i,j \geqslant 0}} \xi_i^{p^j} \otimes \xi_j.$$

*Dans cette formule, par convention $\xi_0 = 1$.*

Ajoutons une graduation en posant : $|\xi_i| = -2(p^i - 1)$. L'application $\delta$ la respecte. On peut alors procéder comme ci-dessus. Cependant (pour $p > 2$) ce résultat ne permet de décrire que la sous-algèbre de l'algèbre de Steenrod engendrée par les puissances réduites, et il ne rend pas compte de l'homomorphisme de Bockstein.

Pour expliquer comment obtenir toute l'algèbre, commençons par revenir au cas $p = 2$. Soit $L$ une $\mathbb{F}_2$-algèbre. On peut décrire $\Gamma(L)$ de la manière suivante. On considère la catégorie $L - \mathcal{A}N^+$ des $L$-algèbres commutatives augmentées sur $L$ dont tout élément de l'idéal d'augmentation est nilpotent. On vérifie facilement :

**Proposition 1.7**. — *Le groupe $\Gamma(L)$ est isomorphe au sous-groupe des équivalences naturelles du foncteur idéal d'augmentation, $\mathcal{I}$, de la catégorie $L - \mathcal{A}N^+$ vers celle des $L$-modules, constitué des équivalences naturelles égales à l'identité sur les algèbres dont tout élément dans l'idéal d'augmentation est de carré nul.*

Cette construction s'étend aussi à $p > 2$, mais encore une fois ceci ne définit que la sous-algèbre de l'algèbre de Steenrod modulo $p$ engendrée par les puissances réduites $\mathrm{P}^i$. Pour obtenir toute l'algèbre $\mathcal{A}_p$, on procède comme suit.

On considère les $\mathbb{F}_p$-algèbres $\mathbb{Z}/2$-graduées commutatives ; la commutativité s'entend ici au sens gradué : $xy = (-1)^{|x||y|}yx$ pour tous $x$, $y$ homogènes ; en particulier les éléments de degré 1 sont de carré nul. Le foncteur oubli $\mathcal{O}$, de cette catégorie vers la catégorie des $\mathbb{F}_p$-algèbres (commutatives ou non), associe à $L$ la somme directe $L^0 \oplus L^1$.

Soit donc $L$ une $\mathbb{F}_p$-algèbre $\mathbb{Z}/2$-graduée commutative. On introduit la sous-catégorie $L - \mathcal{A}N_p^+$ des $L$-algèbres ($\mathbb{Z}/2$-graduées, commutatives) augmentées sur $L$ dont tout élément de l'idéal d'augmentation est nilpotent. Le foncteur idéal d'augmentation, $\mathcal{I}$, est défini de cette catégorie vers la catégorie des $\mathcal{O}(L)$-modules.

**Définition 1.8**. — Le groupe $\widetilde{\Gamma}_p(L)$ est le sous-groupe des équivalences naturelles du foncteur idéal d'augmentation constitué des équivalences naturelles égales à l'identité sur les algèbres dont tout élément dans l'idéal d'augmentation est de carré nul.

**Proposition 1.9**. — *Le groupe $\widetilde{\Gamma}_p(L)$ est produit semi-direct du groupe $\Gamma_p(L^0)$ par le groupe $(L^1)^{\mathbb{N}}$.*

Les conditions imposées montrent que pour une algèbre $M$ dans la catégorie une telle transformation naturelle $\phi$ est donnée par la formule :

$$\phi(x + u) = x + \sum_{i>0} a_i x^{p^i} + \sum_{j \geqslant 0} b_j x^{p^j} + u$$

où $x \in M^0$, $u \in M^1$, $a_i \in L^0$ et $b_j \in L^1$. En effet la restriction d'une équivalence naturelle à la sous-algèbre des éléments de degré 0 détermine un homomorphisme surjectif de $\widetilde{\Gamma}_p(L)$ vers $\Gamma_p(L)$ ; de plus cet homomorphisme est scindé. Enfin $\phi$ est l'identité sur les éléments de degré 1. Pour montrer ceci, il suffit de considérer le cas de l'algèbre extérieure sur $L$ en un générateur $u$ de degré 1 : $u$ est nécessairement envoyé sur lui même. Enfin, on condidère le cas de l'algèbre polynomiale sur $L$ tronquée en un générateur de degré 0, et extérieure en un générateur de degré 1.

En composant deux équivalences naturelles données par :

$$\phi(x + u) = x + \sum_{i>0} a_i x^{p^i} + \sum_{j \geqslant 0} b_j x^{p^j} + u$$

et

$$\psi(y + v) = y + \sum_{i>0} a_i' y^{p^i} + \sum_{j \geqslant 0} b_j' y^{p^j} + v$$

on obtient :

$$\psi \circ \phi(x + u) = x + \sum_{i>0} \left( \sum_{h+k=i} a_h' a_k^{p^h} \right) x^{p^i} + \sum_{j \geqslant 0} \left( \sum_{h+k=j} b_h' a_k^{p^h} \right) x^{p^i} + \sum_{j \geqslant 0} b_j x^{p^j} + u;$$

on reconnaît là une situation analogue à celle du groupe affine.

L'algèbre des fonctions régulières sur ce « schéma en groupes », notée $\mathcal{A}_p^*$, est de la forme

$$\mathbb{F}_p[\xi_i, \ i > 0] \otimes E(\tau_j, \ j \geqslant 0).$$

On gradue cette algèbre sur $\mathbb{Z}$ en posant : $|\xi_i| = -2p^i + 2$ et $|\tau_j| = -2p^j + 1$. On obtient une $\mathbb{Z}/2$-graduation par réduction modulo 2. Comme plus haut, elle admet une structure d'algèbre de Hopf. La diagonale y est donnée sur les éléments $\tau_j$ par :

**Proposition 1.10**

$$\delta(\tau_j) = \tau_j \otimes 1 + \sum_{h+k=j} \xi_h^{p^k} \otimes \tau_k,$$

*où, par convention, $\xi_0 = 1$.*

Cette formule respecte bien la graduation introduite plus haut. L'algèbre graduée duale de $\mathcal{A}_p^*$ est l'algèbre de Steenrod modulo $p$, $\mathcal{A}_p$.

Pour des multi-indices $S = (s_1, \ldots, s_h)$ et $R = (r_1, \ldots, r_h)$, on note $\tau^S \xi^R$ le produit $\tau_1^{s_1} \ldots \tau_h^{s_h} \xi_1^{r_1} \ldots \xi_h^{r_h}$. Une base de $\mathcal{A}_p$ est obtenue par dualisation de la base monomiale de l'algèbre duale $\mathcal{A}_p^*$. :

**Définition 1.11**. — Soit

$$\{\tau^S \xi^R \mid S = (s_1, \ldots, s_h), R = (r_1, \ldots, r_h), h \geqslant 0, 0 \leqslant s_i \leqslant 1, 0 \leqslant r_i, s_h + r_h > 0\}$$

la base monomiale de $\mathcal{A}_p^*$. Sa base duale,

$$\{\mathrm{P}(S, R) = \mathrm{P}(s_1, r_1, \ldots, s_h, r_h) \mid h \geqslant 0, 0 \leqslant s_i \leqslant 1, 0 \leqslant r_i, s_h + r_h > 0\},$$

est appelée base de Milnor de $\mathcal{A}_p$. L'opération $\mathrm{P}(0,i)$ est notée $\mathrm{P}^i$, l'opération $\mathrm{P}(1,0)$ est l'homomorphisme de Bockstein.

**Définition 1.12.** — L'excès de l'opération $\mathrm{P}(s_1, r_1, \ldots, s_h, r_h)$ est l'entier $s_1 + \cdots + s_h + r_1 + \cdots + r_h$.

On note enfin qu'on peut introduire, comme plus haut, des opérations totales $\mathbf{P}$ et $\mathbf{P}^{(i)}$.

## 2. Comodules sur $\mathcal{A}_p^*$, comodules instables, modules instables

**2.1. Comodules et comodules instables.** — Nous étudions maintenant les comodules à droite sur $\mathcal{A}_p^*$. Dans tout ce qui suit, nous traitons les produits infinis qui apparaissent sans soin excessif et nous laissons le lecteur se convaincre que les définitions qui suivent sont sans ambiguités. Les comodules considérés sont gradués, ce sont des $\mathbb{F}_p$-espaces vectoriels $\mathbb{N}$-gradués $M$, munis d'une application de degré 0 :

$$\lambda_M : M \longrightarrow M \,\widehat{\otimes}\, \mathcal{A}_p^*.$$

Le produit tensoriel complété ci-dessus est égal, en degré $n$, au produit

$$\prod_{u+v=n} M^u \otimes (\mathcal{A}_p^*)^v.$$

Rappelons que $\mathcal{A}_p^*$ est gradué négativement.

On note, pour $p = 2$ :

$$\lambda_M(x) = \sum_I x_I \otimes \xi^I,$$

où $I$ désigne un multi-indice, $I = (i_1, \ldots, i_h)$, et $\xi^I = \xi_1^{i_1} \cdots \xi_h^{i_h}$ est un élément de la base monomiale de $\mathcal{A}_2^*$. De manière analogue, pour $p > 2$ :

$$\lambda_M(x) = \sum_I x_I \otimes \xi^I + \sum_I x_{J,K} \otimes \tau^K \xi^J,$$

$I$, $J$ désignant des multi-indices, $I = (i_1, \ldots, i_h)$, $J = (j_1, \ldots, j_\ell)$ et $K$ un multi-indice $(\varepsilon_1, \ldots, \varepsilon_m)$, $\varepsilon_i = 0$ ou $1$. Les éléments $\xi^I$, $\xi^J$, $\tau^K = \tau_1^{\varepsilon_1} \cdots \tau_m^{\varepsilon_m}$, $\xi^J \tau^K$ sont des éléments de la base monomiale de $\mathcal{A}_p^*$. Si, dans la formule, $x$ est un élément homogène de degré $n$, le terme de droite, composante par composante, est homogène de même degré. L'application $\lambda_M$ doit évidemment satisfaire aux identités duales de celles définissant une structure de module.

**Définition 2.1.** — Pour $p = 2$, un comodule $M$ est instable si les facteurs $x_I$ sont nuls dès que $e(I) = i_1 + \cdots + i_h > n$.

**Définition 2.2.** — Pour $p > 2$, un comodule $M$ est instable si les facteurs $x_I$, respectivement $x_{J,K}$, sont nuls dès que $i_1 + \cdots + i_h > n$, respectivement $j_1 + \cdots + j_\ell + \varepsilon_1 + \cdots + \varepsilon_m > n$.

Une telle structure est équivalente à une « représentation à droite du schéma affine en groupes $\Gamma(-)$ ». Une telle représentation est, par définition, la donnée d'un $\mathbb{F}_p$-espace vectoriel $\mathbb{N}$-gradué $M$ et, pour toute $\mathbb{F}_p$-algèbre $\mathbb{N}$-graduée, $A$ d'un homomorphisme naturel en $A$ :

$$\Gamma(A) \longrightarrow \mathrm{GL}(A \widehat{\otimes} M)$$

(on doit ici encore compléter le produit tensoriel).

Par dualité, une structure de comodule à droite sur $M$ induit une structure de $\mathcal{A}_p$-module à gauche. L'action de l'algèbre de Steenrod est donnée par :

$$\theta x = \langle \lambda_M(x), \theta \rangle = \sum_I x_I \langle \xi_I, \theta \rangle,$$

avec les notations correspondant à $p = 2$ pour la seconde égalité, où $\langle \, , \, \rangle$ désigne le produit de Kronecker. On peut donc reécrire l'application de structure du comodule comme suit :

$$\lambda_M(x) = \sum_I \mathrm{Sq}(I) x \otimes \xi^I,$$

et de manière analogue pour $p > 2$. Si le comodule est instable, on dira que le module l'est. Les modules instables vérifient les propriétés suivantes.

**Corollaire 2.3**. — *Pour $p = 2$, si $x$ est dans un module instable : $\mathrm{Sq}^i(x) = 0$ dès que $i > |x|$.*

**Corollaire 2.4**. — *Pour $p > 2$, si $x$ est dans un module instable : $\mathrm{P}^i(x) = 0$ dès que $2i > |x|$ et $\beta \mathrm{P}^i(x) = 0$ si $2i + 1 > |x|$.*

En fait, ainsi qu'on le verra ci-dessous (proposition 2.5), ces conditions sont nécessaires et suffisantes.

**2.2. Exemples fondamentaux et propriétés.** — Voici un exemple fondamental de comodule instable. Considérons le sous-espace vectoriel gradué $L$ de $\mathcal{A}_2^*$ engendré par $1$ et les $\xi_i$, $i > 0$. La formule pour la diagonale $\delta$ montre que : $\delta(L) \subset \mathcal{A}_2^* \otimes L$. Cette formule ne respecte pas la convention de degré imposée plus haut ; de plus elle définit un objet à gauche. Pour contourner cette difficulté, considérons la suspension du dual de $L$, que l'on notera $F(1)$. On a donc : $F(1)^n = (L_{-n+1})^{\#}$, et l'application $F(1) \to F(1) \otimes \mathcal{A}_2^*$ est obtenue par dualisation partielle de $L \to \mathcal{A}_2^* \otimes L$. L'espace vectoriel gradué $F(1)$ est de dimension 1 dans les degrés $2^i$, on en note $v_i$ le générateur, il est nul dans les autres degrés. C'est un comodule à droite sur $\mathcal{A}_2^*$ : sa structure est donnée par la formule :

$$\lambda_{F(1)}(v_i) = v_i + \sum_{j>0} v_{i+j} \otimes \xi_j^{2^i}.$$

Le calcul s'effectue en utilisant la remarque suivante. Si $x \in F(1)$ et si $\lambda_{F(1)}(x) = \sum_I x_I \otimes \xi^I$, $\ell \in L$ et $\delta(\ell) = \sum_I \xi^I \otimes \ell_I$, alors :

$$x_I(\ell) = x(\ell_I).$$

En particulier pour $x = v_0 \in F(1)^1$, on a :

$$\langle \theta(v_0), \xi_i \rangle = \langle v_0, \langle \theta, \delta(\xi_i) \rangle \rangle$$

et $\langle \theta, \delta(\xi_i) \rangle = \sum_{k+h=i} \langle \theta, \xi_h^{p^k} \rangle \xi_k$. Il en résulte que le module $F(1)$ est monogène. En effet, pour $\Delta_i = (0, \ldots, 1, \ldots)$, où le terme 1 est en $i$-ième position, on a : $\mathrm{Sq}(\Delta_i)v_0 = v_i$. L'action des opérations de Steenrod est donc donnée par : $\mathrm{Sq}^{2^i}(v_i) = v_{i+1}$, et $\mathrm{Sq}^{2^i}(v_j) = 0$ sinon.

Le générateur de $F(1)^1$ s'interprète en topologie comme le générateur de la cohomologie modulo 2 du groupe $\mathbb{Z}/2\mathbb{Z}$. Suivant l'usage, on note $u$ cet élément.

De manière analogue pour $p > 2$, on définit les modules instables. On considère le sous-module de $\mathcal{A}_p^*$ engendré par les $\xi_i$ et l'unité. On le dualise, et en suspendant deux fois, on obtient un module que l'on note $F'(2)$.

Encore une fois, un calcul direct montre que, si $w_i$ désigne un générateur de $F'(2)$ en degré $2p^i$ (ils sont nuls dans les autres degrés), on a :

$$\lambda_{F'(2)}(w_i) = w_i + \sum_{j>0} w_{i+j} \otimes \xi_j^{p^i}.$$

L'action des opérations de Steenrod est donc donnée par : $\mathrm{P}^{p^i}(w_i) = w_{i+1}$ et $\mathrm{P}^{p^i}(w_j) = 0$ sinon.

Toujours pour $p > 2$, on construit un autre module en considérant le sous-module de $\mathcal{A}_p^*$ engendré par les $\tau_i$ et l'unité. On le dualise, et en suspendant une fois on obtient un module que l'on note $F(1)$. Soit $t$ le générateur en degré 1 et $x_i$ les générateurs de $F(1)$ en degré $2p^i$ (ils sont nuls dans les autres degrés). On a :

$$\lambda_{F(1)}(t) = t + \sum_{j \geqslant 0} x_j \otimes \tau_j,$$
$$\lambda_{F(1)}(x_i) = x_i + \sum_{j>0} x_{i+j} \otimes \xi_j^{p^i}.$$

L'action des opérations de Steenrod est donc donnée par $\mathrm{P}^{p^i}(x_i) = x_{i+1}$, $\beta(t) = x_0$ et $\mathrm{P}^{p^i}(x_j) = 0$ sinon.

Pour les mêmes raisons que plus haut, ces deux modules sont monogènes, et $F'(2)$ est de dimension 1 dans les degrés $2p^i$, trivial sinon ; $F(1)$ est de dimension 1 dans les degrés $2p^i$ et 1, trivial sinon.

Comme cet élément s'interprète comme le générateur en degré 2 de la cohomologie modulo $p$ du groupe $\mathbb{Z}/p\mathbb{Z}$, on rebaptise $x$ le générateur de $F'(2)^2$.

C'est une conséquence des relations d'Adem, démontrées plus loin, que la réciproque des corollaires 2.3 et 2.4 est vraie :

**Proposition 2.5**. — *Un module est instable si, et seulement si, il vérifie les conditions des corollaires 2.3 ou 2.4.*

La démonstration de cette proposition est laissée en exercice ; elle repose essentiellement sur la construction de la base des monômes de Cartan-Serre, elle-même conséquence des relations d'Adem.

La catégorie des modules instables, dont les morphismes sont les applications $\mathcal{A}_p$-linéaires de degré 0, est notée $\mathcal{U}$. C'est une catégorie abélienne.

## 2.3. Produit tensoriel de (co)-modules, opérations multiplicatives

Étant donnés deux comodules $M$ et $N$, on définit une structure de comodule sur leur produit tensoriel $M \otimes N$ par la formule ($p = 2$) :

$$\lambda_{M \otimes N}(x \otimes y) = \sum_K \left( \sum_{I+J=K} x_I \otimes y_J \right) \otimes \xi^K,$$

où l'on a $\lambda_M(x) = \sum_I x_I \otimes \xi^I$ et $\lambda_N(y) = \sum_J y_J \otimes \xi^J$.

On étend à $p > 2$, en prenant garde aux signes :

$$M \otimes N \xrightarrow{\lambda_M \otimes \lambda_N} M \otimes \mathcal{A}_p^* \otimes N \otimes \mathcal{A}_p^*$$

$$\xrightarrow{\mathrm{Id} \otimes T \otimes \mathrm{Id}} M \otimes N \otimes \mathcal{A}_p^* \otimes \mathcal{A}_p^* \xrightarrow{\mathrm{Id} \otimes \mathrm{Id} \otimes \mathrm{mult}} M \otimes N \otimes \mathcal{A}_p^*,$$

avec $T(x \otimes y) = (-1)^{|x|\,|y|}$.

En particulier, on définit la suspension d'un (co)-module (la condition d'instablité n'est ici pas nécessaire) par :

$$\Sigma M = \Sigma \mathbb{F}_p \otimes M,$$

où $\Sigma \mathbb{F}_p$ est le (co)-module (instable) égal à $\mathbb{F}_p$ en dimension 1, $\{0\}$ sinon, muni de l'action triviale. On a :

– $(\Sigma M)^n \cong M^{n-1}$,

– l'action de $\mathcal{A}_p$ sur $\Sigma M$ est donnée par : $\theta(\Sigma m) = (-1)^{|\theta|} \Sigma \theta m$ pour tout élément $m \in M$ et toute opération $\theta \in \mathcal{A}_p$.

Le résultat suivant est immédiat.

**Lemme 2.6**. — *Le produit tensoriel de deux comodules instables est un comodule instable. Il en est de même des modules associés. La suspension d'un module instable est instable.*

Si $M$ est un comodule, un élément $\gamma \in \Gamma(\mathbb{F}_p) \subset \prod_i \mathcal{A}_p^i$ induit une application $\widehat{\gamma}$ de $\widehat{M} = \prod_i M^i$ dans lui même. Pour un élément homogène $\gamma$, elle est donné par la formule :

$$\widehat{\gamma}(m) = \langle \lambda_M(m), \gamma \rangle = \sum_I \gamma(\xi^I)\, m_I$$

On étend formellement $\widehat{\gamma}$ à un élément du produit $\widehat{M} = \prod_i M^i$, car, en un degré donné, seuls un nombre fini d'éléments donnent une contribution non nulle.

Plus généralement, si $L$ est une $\mathbb{F}_2$-algèbre (non graduée) et $M$ un comodule, un élément $\gamma \in \Gamma(L) \subset \prod_i \mathcal{A}_p^i \otimes L$ induit une application $\widehat{\gamma}$ de $\widehat{M \otimes L} = \prod_i M^i \otimes L$ dans lui même. Sur un élément homogène, $\widehat{\gamma}$ est donné par la même formule :

$$\widehat{\gamma}(m) = \sum_I \gamma(\xi^I) \, m_I$$

Le résultat suivant est une conséquence de la définition :

**Lemme 2.7**. — *L'application $\widehat{\gamma}$ est multiplicative : pour $x \in M$ et $y \in N$,*

$$\widehat{\gamma}(x \otimes y) = \widehat{\gamma}(x) \otimes \widehat{\gamma}(y).$$

Rappelons :

**Définition 2.8**. — Si $\gamma \in \Gamma(L)$ prend la valeur $\ell \in L$ sur $\xi_1$ et $0$ sur $\xi_i$, $i > 1$, on note alors l'opération $\widehat{\gamma} : \widehat{M} \to \widehat{M}$ par $\mathbf{Sq}^{(1)}(\ell)$, ou plus communément $\mathbf{Sq}(\ell)$.

## 3. Les (co)modules instables libres

La catégorie $\mathcal{U}$ des modules instables est abélienne. On va montrer qu'elle a assez d'objets projectifs. Les théorèmes de Cartan-Serre contiennent implicitement ce résultat. La description des objets libres de $\mathcal{U}$ a été explicitée et développée par Massey et Peterson [**18**]. Nous suivons ici la présentation des sections précédentes.

Dans le cas $p = 2$, affectons à chaque $\xi_i$ le poids 1 ; dans le cas $p > 2$, affectons à chaque $\xi_i$ le poids 2, et à chaque $\tau_i$ le poids 1. Le poids d'un monôme $\xi^I$ est alors $\sum_\alpha i_\alpha$ pour $p = 2$, $\sum_\alpha 2 i_\alpha$ pour $p > 2$, et celui d'un monôme $\xi^J \tau^K$, $J = (j_1, \ldots, j_\ell)$, $K = (\varepsilon_1, \ldots, \varepsilon_m)$, est : $\sum_\alpha 2 j_\alpha + \sum_\beta \varepsilon_\beta$.

Pour $p = 2$, considérons le sous-objet $P_h \cong S^h(L)$ ($L$ est défini en section 2.2) des monômes de poids inférieur ou égal à $h \geqslant 0$ dans $\mathcal{A}_2^*$. La diagonale $\delta$ envoie $P_h$ dans $\mathcal{A}_2^* \otimes P_h$. Comme plus haut, pour des raisons de degré, ceci ne définit pas un comodule. Mais comme plus haut aussi, par dualisation partielle et suspension $h$-fois, on obtient un objet $F(h)$ (avec $F(h)^k = (P_h^{k-h})^\#$), qui est un comodule instable à droite. Le groupe symétrique $\mathfrak{S}_h$ agit sur le module instable $F(1)^{\otimes h}$ par permutation des facteurs. Par construction, on a :

**Proposition 3.1**. — *Pour $p = 2$ :*

$$F(h) \cong \Gamma^h(F(1)) \cong (F(1)^{\otimes h})^{\mathfrak{S}_h}.$$

*Ce module instable est engendré par son unique classe non nulle $\iota_h$ de degré $h$.*

La première partie résulte de l'isomorphisme naturel rappelé en introduction. La seconde résulte de la construction. Par définition, on a en effet :

$$\lambda_{F(h)}(\iota_h) = \sum_I (\iota_h)_I \otimes \xi^I,$$

ainsi que la relation suivante :

$$\langle (\iota_h)_I, \xi^J \rangle = \langle \iota_h, \alpha_I^J \rangle$$

où $\delta(\xi_J) = \sum_I \alpha_I^J \otimes \xi_I$. Les classes $(\iota_h)_I$ sont donc les classes duales des classes $\xi^I$ de poids inférieur ou égal à $h$. En particulier, on a : $\mathrm{Sq}(I)\iota_h = (\iota_h)_I$. Les classes $(\iota_h)_I$ forment donc une base de $F(h)$, par construction, et sont linéairement indépendantes : le module $F(h)$ est monogène.

Considérons maintenant le cas $p > 2$, et le sous-objet $P_h$ de $\mathcal{A}_p^*$ constitué par les sommes de monômes de poids inférieur ou égal à $h$. Comme précédemment, on a : $\delta(P_h) \subset \mathcal{A}_p^* \otimes P_h$. Désignons par $L_1$ le sous-espace engendré par les $\tau_i$ et l'unité, et par $L_2$ le sous-espace engendré par les $\xi_j$ et l'unité. Le sous-comodule $P_h$ s'identifie à $S^h(L_1 \oplus L_2)$. Le foncteur $S^h$ est dans ce contexte compris en un sens gradué ; on suppose les éléments de $L_1$ affectés du degré 1, ceux de $L_2$ affectés du degré 0. On considère la somme directe

$$\bigoplus_{2\ell+k=h} L_2^{\otimes \ell} \otimes L_1^{\otimes k}$$

et on quotiente par les relations engendrées par $x \otimes y = (-1)^{|x|\,|y|} y \otimes x$.

On démontre, comme dans le cas $p = 2$, que :

**Proposition 3.2**. — *Le dual gradué de $P_h$, suspendu $h$-fois, que l'on note $F(h)$, est monogène en générateur $\iota_h$ de degré $h$. Les classes $\mathrm{P}(J,K)(\iota_h)$, où $\mathrm{P}(J,K)$ est d'excès inférieur ou égal à $h$, en forment une base.*

On n'a pas dans ce cas d'identification de $F(h)$ avec les puissances divisées. Ceci étant, on a :

**Proposition 3.3**. — *Le module $F(h)$ admet une filtration croissante dont les quotients successifs sont les modules instables*

$$\Gamma^\ell(F'(2)) \otimes \Lambda^k(F(1)),$$

*avec $2\ell + k = h$. Si $h$ est pair, le module quotient $\Gamma^{h/2}(F'(2))$ est noté $F'(2h)$. C'est le module instable libre, avec action triviale de l'homomorphisme de Bockstein, en un générateur de degré $h$.*

**Théorème 3.4**. — *La transformation naturelle*

$$f \longmapsto f(\iota_h) \quad \mathrm{Hom}_\mathcal{U}(F(h), M) \longrightarrow M^h$$

*est une équivalence de foncteurs.*

Ce (co)-module est libre en ce sens qu'il n'y a aucune relation de plus que celle directement impliquée par la condition d'instabilité : $\mathrm{Sq}(I)\iota_h$ est nul si, et seulement si, l'excès de $I$ est strictement supérieur à $h$, et les classes $\mathrm{Sq}(I)\iota_h$, où l'excès de $I$ est inférieur à $h$, sont linéairement indépendantes. Le résultat suit.

Le module $F(h)$ est appelé module instable libre en un générateur de degré $h$.

## 4. Les relations d'Adem

**4.1. Le théorème de Bullett et MacDonald.** — Le but de cette section est d'établir les relations d'Adem, qui sont duales de la diagonale $\delta$, puis de décrire la base de Cartan-Serre de l'algèbre de Steenrod.

On va procéder en suivant l'approche de Bullett et MacDonald. Soit $L$ une $\mathbb{F}_2$-algèbre (resp. une $\mathbb{F}_p$-algèbre $\mathbb{Z}/2$-graduée). On considère des éléments $\gamma_i$, $i = 1, 2, 3, 4$, appartenant au groupe $\Gamma(L)$ (resp. $\widetilde{\Gamma}_p(L)$). Ces éléments induisent (voir après le lemme 2.6), pour tout module instable $M$, une application $\widehat{\gamma}$ de $\prod_i M^i \otimes L$ dans lui même. Supposons que : $\gamma_1 \circ \gamma_2(u) = \gamma_3 \circ \gamma_4(u)$, où $u$ désigne le générateur de $F(1)^1$, que l'on soit en en $p = 2$ ou en $p > 2$.

**Proposition 4.1.** — *Si $\gamma_1(\gamma_2(u)) = \gamma_3(\gamma_4(u))$, alors $\gamma_1 \circ \gamma_2 = \gamma_3 \circ \gamma_4$.*

En effet, les opérations considérées sont multiplicatives (lemme 2.7). Si donc $\gamma_1 \circ \gamma_2$ et $\gamma_3 \circ \gamma_4$ coïncident sur $u$, elles coïncident sur les produits tensoriels de $F(1)$, donc sur tout sous-objet ou quotient, donc sur les modules $F(h)$ pour tout $h$. Par définition de l'action d'une opération $\widehat{\gamma}$ et des modules $F(h)$, les éléments $\gamma_1 \circ \gamma_2$ et $\gamma_3 \circ \gamma_4$ sont égaux.

Prenons pour $L$ l'algèbre $\mathbb{F}_2[s, t]$ polynomiale en deux générateurs.

**Théorème 4.2.** — *On a la formule suivante* [3] :
$$\mathbf{Sq}(st + t^2)\mathbf{Sq}(s^2) = \mathbf{Sq}(st + s^2)\mathbf{Sq}(t^2).$$

Pour vérifier cette formule, il suffit de le faire sur la classe $u$. Or : $\mathbf{Sq}(\ell)(u) = u + \ell u^2$. On obtient alors :
$$\mathbf{Sq}(st + t^2)\mathbf{Sq}(s^2)(u) = \mathbf{Sq}(st + t^2)(u + s^2 u^2) = u + (st + s^2 + t^2)u^2 + s^2 t^2 (s^2 + t^2)u^4,$$

qui est symétrique en $s$ et $t$. On fait $s = 1$, on a :
$$\mathbf{Sq}(t + t^2)\mathbf{Sq}(1) = \mathbf{Sq}(t + 1)\mathbf{Sq}(t^2).$$

Posons $\tau = t + t^2$. Le terme de gauche de l'égalité ci-dessus est la série formelle en $\tau$ suivante :
$$\sum_{a \geqslant 0} \Big(\sum_{b \geqslant 0} \mathrm{Sq}^a \mathrm{Sq}^b\Big) \tau^a.$$

Dans cette formule, le coefficient de $\tau^a$ est un élément du produit $\prod_i \mathcal{A}_2^i$, et il est écrit abusivement comme une somme. La variable $t$ s'exprime comme série formelle en $\tau$, $t = \varphi(\tau) = \tau + \tau^2 + \tau^4 + \cdots$. Si on exprime le terme de droite comme série formelle en $\tau$, le coefficient de $\tau^a$ dans cette série est le résidu :
$$\mathrm{Res}_{\tau=0} \, \mathbf{Sq}(\varphi(\tau) + 1)\mathbf{Sq}(\varphi(\tau)^2)\frac{d\tau}{\tau^{a+1}}.$$

Faisons le changement de variables de $\tau$ en $t$ et exprimons le coefficient. La formule de changement est analogue à celle dans une intégrale et donc dans le théorème des résidus : comme $d\tau = dt$, le coefficient cherché est :

$$\text{Res}_{t=0} \, \mathbf{Sq}(t+1)\mathbf{Sq}(t^2)\frac{dt}{t^{a+1}(1+t)^{a+1}},$$

soit le coefficient de $t^a$ dans :

$$\sum_{a,b,j \geqslant 0} (1+t)^{b-j-1}t^{2j}\mathrm{Sq}^{a+b-j}\mathrm{Sq}^{j}.$$

Il suffit alors d'identifier les composantes des coefficients degré par degré pour obtenir les relations d'Adem ci-dessous :

$$\mathrm{Sq}^a\mathrm{Sq}^b = \sum_{0}^{[a/2]} \binom{b-j-1}{a-2j}\mathrm{Sq}^{a+b-j}\mathrm{Sq}^{j}$$

pour tous $a, b > 0$. Notons que la restriction $a < 2b$ n'est pas utile.

Pour $p > 2$, on raisonne de manière analogue en introduisant l'opération $\mathbf{P}(\ell)$, ce qui donne la relation :

$$\mathbf{P}(\tau)\mathbf{P}(1) = \mathbf{P}(u)\mathbf{P}(t^p)$$

avec $u = 1 + \cdots + t^{p-1}$ et $\tau = tu$. Le même procédé que plus haut donne les relations suivantes :

$$\mathrm{P}^a\mathrm{P}^b = \sum_{0}^{[a/p]} (-1)^{a+j}\binom{(p-1)(b-j)-1}{a-pj}\mathrm{P}^{a+b-j}\mathrm{P}^{j}$$

pour tous $a, b > 0$ (la condition $a < pb$ n'est pas utile).

Les relations incluant l'homomorphisme de Bockstein sont obtenues comme suit. On considère l'opération $\mathbf{P}(t) + v[\beta, \mathbf{P}(t)]$, où $v$ est un élément de degré 1 de carré nul. On montre facilement que cette opération est multiplicative. On pourra vérifier qu'elle correspond, en tant qu'élément du groupe $\Gamma_p(\mathbb{F}_p[t] \otimes E(v))$, à l'homomorphisme qui envoie $\zeta_1$ sur $t$, les autres $\xi_i$ sur 0, $\tau_1$ sur $v$ et les autres $\tau_i$ sur 0. Par ailleurs, on a :

$$(\mathbf{P}(\tau) + v[\beta, \mathbf{P}(\tau)])\mathbf{P}(1) = \mathbf{P}(u) + vt[\beta, \mathbf{P}(u)]\mathbf{P}(t^p).$$

Un calcul analogue aux précédents donne alors :

$$\mathrm{P}^a\beta\mathrm{P}^b - \sum_{0}^{[a/p]} (-1)^{a+i}\binom{(p-1)(b-i)}{a-pt}\beta\mathrm{P}^{a+b-i}\mathrm{P}^{i}$$

$$- \sum_{0}^{[(a-1)/p]} (-1)^{a+i-1}\binom{(p-1)(b-i)-1}{a-pi-1}\mathrm{P}^{a+b-i}\beta\mathrm{P}^{t}$$

pour tous $a, b > 0$.

D'autre choix des $\gamma_i$ donneraient d'autres relations.

**4.2. La base de Cartan-Serre.** — Voici une conséquence facile des relations. Soit $I = (i_1, i_2, \ldots, i_n)$ une suite d'entiers, elle sera dite *admissible* si $2i_\alpha \geqslant i_{\alpha+1}$ pour tout $\alpha$. L'excès de cette suite $(i_1, i_2, \ldots, i_n)$ est défini par la formule :

$$(i_1 - 2i_2) + (i_2 - 2i_3) + \cdots + (i_{n-1} - 2i_n) + i_n.$$

À $I = (i_1, i_2, \ldots, i_n)$, on associe l'opération $\mathrm{Sq}^{i_1} \cdots \mathrm{Sq}^{i_n}$, que l'on note $\mathrm{Sq}^I$. Cette opération est appelée *monôme admissible* de Cartan-Serre.

Pour $p > 2$, on considère des suites $I = (\varepsilon_0, i_1, \varepsilon_1, i_2, \ldots, i_n, \varepsilon_n)$, l'excès est donné par :

$$2(i_1 - pi_2) + 2(i_2 - pi_3) + \cdots + 2i_n + \varepsilon_0 - \varepsilon_1 - \cdots - \varepsilon_n,$$

les $\varepsilon_i$ valant 0 ou 1. L'excès est noté $e(I)$. On associe l'opération $\mathrm{P}^I = \beta^{\varepsilon_0} \mathrm{P}^{i_1} \cdots \beta^{\varepsilon_n}$ appelée monôme admissible de Cartan-Serre.

Il y a donc deux notions d'excès, l'une pour les monômes de Cartan-Serre, et l'autre pour les opérations de Milnor. Ces deux notions sont compatibles.

**Théorème 4.3.** — *Les opérations admissibles $\mathrm{Sq}^I$ (resp. $\mathrm{P}^I$) forment une base de $\mathcal{A}_p$.*

Les relations d'Adem montrent que c'est un système générateur. Un argument de série de Poincaré permet de conclure.

Donnons une autre description du module $F(h)$. Soit $\iota_h$ le générateur de $F(h)$ en degré $h$.

**Théorème 4.4.** — *Le module instable $F(h)$ est isomorphe à*

$$\Sigma^h \big( \mathcal{A}_p / (\mathrm{P}^I, I \text{ admissible}, e(I) > h) \big).$$

Les lemmes 2.3 et 2.4 montrent que $\mathrm{Sq}^I \iota_h$ et $\mathrm{P}^I \iota_h$ sont nuls si $e(I) > h$. Ce module a pour base, comme $\mathbb{F}_p$-espace vectoriel gradué, les éléments $\Sigma^h \mathrm{P}^I$, avec $I$ admissible et $e(I) \leqslant h$. On note $\iota_n$ pour $\Sigma^h \mathrm{P}^0$. On montre que le $\mathbb{F}_p$-espace vectoriel gradué engendré par les $\Sigma^n \mathrm{P}^I$, avec $I$ admissible tel que $e(I) > h$, est un sous-$\mathcal{A}_p$-module de $\mathcal{A}_p$.

**4.3. Générateurs multiplicatifs.** — Voici une autre application des relations d'Adem.

**Théorème 4.5.** — *Pour $p = 2$, les opérations $\mathrm{Sq}^{2^h}$, $h \geqslant 0$, constituent un système de générateurs multiplicatifs de $\mathcal{A}_2$. Pour $p > 2$, ce sont les opérations $\beta$ et $\mathrm{P}^{p^h}$, $h \geqslant 0$, qui jouent ce rôle pour $\mathcal{A}_p$.*

Cela résulte du lemme suivant qui se démontre à partir des relations d'Adem :

**Lemme 4.6.** — *Pour $p = 2$ (resp. $p > 2$), l'opération $\mathrm{Sq}^m$ (resp. $\mathrm{P}^m$) appartient à l'idéal à droite de $\mathcal{A}_p$ engendré par les $\mathrm{Sq}^{2^t}$ (resp. $\mathrm{P}^{p^t}$) tels que $t \geqslant 0$ de $2^t$ (resp. $p^t$) divise $m$.*

## 5. Retour sur la condition d'instabilité

**5.1. Algèbres instables.** — Nous rappelons quelques points de topologie, repris aussi dans l'article annexe. La cohomologie modulo $p$ d'un espace $X$ est un $\mathcal{A}_p$-module instable, au sens suivant. Pour toute classe de cohomologie $x$, de degré $|x|$ :

- pour $p = 2$, si $i > |x|$, alors $\mathrm{Sq}^i x = 0$ ;
- pour $p > 2$, si $e + 2i > |x|$, $e = 0$ ou $1$, alors $\beta^e P^i x = 0$.

**Définition 5.1.** — Un $\mathcal{A}_p$-module $M$ est dit instable si il satisfait la condition précédente.

La notion d'instabilité n'a pas été introduite ci-dessus à partir de cette condition, mais celle ci est équivalente à la condition des définitions 2.1 et 2.2.

La cohomologie modulo $p$ d'un espace $X$ est, de plus, munie du cup-produit, une $\mathbb{F}_p$-algèbre $\mathbb{N}$-graduée, commutative, unitaire, et ce naturellement. Cette structure est reliée à la structure de $\mathcal{A}_p$-module par les deux propriétés de la définition qui suit.

**Définition 5.2.** — Une $\mathcal{A}_p$-algèbre instable $K$ est un module instable muni de deux applications $\mu : K \otimes K \to K$ et $\eta : \mathbb{F}_p \to K$ qui déterminent une structure de $\mathbb{F}_p$-algèbre commutative, unitaire, vérifiant les propriétés suivantes.

$(\mathcal{K}_1)$ la « formule de Cartan » :
- pour $p = 2$ :

$$\mathrm{Sq}^i(xy) = \sum_{k+\ell=i} \mathrm{Sq}^k x \, \mathrm{Sq}^\ell y;$$

- pour $p > 2$ :

$$P^i(xy) = \sum_{k+\ell=i} P^k x \, P^\ell y,$$

$$\beta(xy) = (\beta x)\, y + (-1)^{|x|} x \, \beta y.$$

pour tous $x, y \in K$.

$(\mathcal{K}_2)$
- pour $p = 2$ : $\mathrm{Sq}^{|x|} x = x^2$, pour tout $x \in K$ ;
- pour $p > 2$ : $P^{|x|/2} x = x^p$, pour tout $x \in K$ de degré pair.

Rappelons les exemples fondamentaux présentés dans l'article annexe.

**Exemple 5.3.** — La cohomologie modulo 2 du groupe $\mathbb{Z}/2$, $\mathrm{H}^*(\mathbb{Z}/2)$, est une algèbre polynomiale $\mathbb{F}_2[u]$ en un générateur $u$ de degré 1. L'action de $\mathcal{A}_2$ est complètement déterminée par $\mathcal{K}_1$ et $\mathcal{K}_2$. On trouve :

$$\mathrm{Sq}^i u^n = \binom{n}{i} u^{n+i}.$$

Pour $p > 2$, $\mathrm{H}^*(\mathbb{Z}/p)$ est le produit tensoriel $E(t) \otimes \mathbb{F}_p[x]$ d'une algèbre extérieure en un générateur $t$ de degré 1 et d'une algèbre polynomiale en un générateur $x$ de degré 2.

L'action de $\mathcal{A}_p$ est determinée par $\mathcal{K}_1$, $\mathcal{K}_2$ et le fait que $\beta$ est l'homomorphisme de Bockstein. On obtient :

$$\beta t = x, \quad P^i x^n = \binom{n}{i} x^{n+i(p-1)}.$$

**Exemple 5.4**. — Voici un exemple plus général. Soit $V$ un p-groupe abélien élémentaire, c'est-à-dire un espace vectoriel de dimension finie sur $\mathbb{F}_p$. La cohomologie modulo $p$ de l'espace classifiant $BV$, ou cohomologie du groupe $V$, est notée $\mathrm{H}^*V$.

Pour $p = 2$, $\mathrm{H}^*V$ s'identifie à l'algèbre symétrique, $S^*(V^\#)$, engendrée par $V^\#$ concentré en degré 1. C'est-à-dire à l'algèbre polynomiale en $d$ générateurs $x_i$ de degré 1, si $d$ est la dimension de $V$.

Pour $p > 2$, $\mathrm{H}^*V$ s'identifie au produit tensoriel de l'algèbre extérieure, $E(V^\#)$, engendrée par $V^\#$ concentré en degré 1, par l'algèbre symétrique, $S^*(V^\#)$, engendrée par $V^\#$ concentré en degré 2. C'est-à-dire, si $d$ est la dimension de $V$, le produit tensoriel de l'algèbre polynomiale en $d$ générateurs $x_i$ de degré 2 par l'algèbre extérieure en $d$ générateurs $x_i$ de degré 1.

Pour $p = 2$, une application d'algèbres de $\mathrm{H}^*W$ dans $\mathrm{H}^*V$ est déterminée par ses valeurs sur les éléments de degré 1. Dans ce degré, elle s'identifie à une application linéaire $\varphi : V \to W$. L'application d'algèbres $\phi^*$ induite par $\phi$ est $\mathcal{A}_p$-linéaire. L'action de $\mathcal{A}_2$ est donnée par les formules ci-dessus. Il en résulte facilement que :

$$\mathrm{Hom}(V, W) \longrightarrow \mathrm{Hom}_{\mathcal{K}}(\mathrm{H}^*W, \mathrm{H}^*V), \quad \varphi \longmapsto \varphi^*,$$

est une bijection. La même observation vaut pour $p > 2$.

**Exemple 5.5**. — Revenons sur les deux exemples initiaux de module instable qui sont fondamentaux. On observe que :

– Pour $p = 2$, $F(1)$ s'identifie au sous-$\mathcal{A}_2$-module de $\mathrm{H}^*\mathbb{Z}/2$ engendré par $u$. L'ensemble $\{u, u^2, u^4, \ldots\}$ est une base de $F(1)$ comme $\mathbb{F}_2$-espace vectoriel gradué.

– Pour $p > 2$, $F(1)$ s'identifie au sous-$\mathcal{A}_p$-module de $\mathrm{H}^*\mathbb{Z}/p$ engendré par $t$. L'ensemble $\{t, x, x^p, x^{p^2}, \ldots\}$ est une de base $F(1)$ comme $\mathbb{F}_p$-espace vectoriel gradué.

## 5.2. Les opérations $\mathrm{Sq}_0$ et $\mathrm{P}_0$.

— Certaine relations d'Adem se simplifient dans le contexte des modules instables. On définit l'opération $\mathrm{Sq}_0$, pour $p = 2$, par :

$$\mathrm{Sq}_0 x = \mathrm{Sq}^{|x|} x,$$

et l'opération $\mathrm{P}_0$, pour $p > 2$, par :

$$\mathrm{P}_0 x = \mathrm{P}^{|x|/2} x \text{ si } x \text{ est de degré pair,}$$

et

$$\mathrm{P}_0 x = \beta \mathrm{P}^{(|x|-1)/2} x \text{ si } x \text{ est de degré impair.}$$

On vérifie que, pour $x$ dans un module instable, pour tout $i \geqslant 0$, on a :

– pour $p = 2$ :

$$\mathrm{Sq}^i\mathrm{Sq}_0 x = \begin{cases} \mathrm{Sq}_0\mathrm{Sq}^{i/2}x & \text{si } i \equiv 0 \ (2), \\ 0 & \text{si } i \equiv 1 \ (2); \end{cases}$$

– pour $p > 2$ :

$$\mathrm{P}^i\mathrm{P}_0 x = \begin{cases} \mathrm{P}_0\mathrm{P}^{i/p}x & \text{si } i \equiv 0 \ (p), \\ \mathrm{P}_0\beta\mathrm{P}^{(i-1)/p}x & \text{si } i \equiv 1 \ (p) \text{ et } |x| \equiv 1 \ (2), \\ 0 & \text{sinon}. \end{cases}$$

Il résulte des formules ci-dessus que le sous-objet $\mathrm{Sq}_0(F(n)) \subset F(n)$ (resp. $\mathrm{P}_0(F(n))$) est un sous-module instable. Une base de ce sous-module est donnée par les classes $\mathrm{Sq}^I\iota_n$, avec $I$ admissible et d'excès égal à $n$. On a aussi une application canonique : $F(n) \to \Sigma F(n-1)$, déterminée par un générateur de $\Sigma F(n-1)$ en degré $n$. Cette application est nulle sur $\mathrm{Sq}_0(F(n))$. L'examen des bases de Cartan-Serre donne alors la suite exacte :

$$0 \longrightarrow \mathrm{Sq}_0(F(n)) \longrightarrow F(n) \longrightarrow \Sigma F(n-1) \longrightarrow 0.$$

On a évidemment une suite exacte analogue pour $p > 2$ :

$$0 \longrightarrow \mathrm{P}_0(F(n)) \longrightarrow F(n) \longrightarrow \Sigma F(n-1) \longrightarrow 0,$$

avec les propriétés analogues concernant les bases.

## 6. Sur la structure de la catégorie $\mathcal{U}$

On donne ici quelques compléments sur la structure de $\mathcal{U}$ (voir [**21**]).

**6.1. Modules instables nilpotents, réduits et localement finis.** — On définit certains types de modules instables. Commençons par rappeler qu'une sous-catégorie pleine d'une catégorie abélienne est *épaisse* dès que, étant donnée une suite exacte :

$$0 \longrightarrow M' \longrightarrow M \longrightarrow M'' \longrightarrow 0$$

dont deux des objets sont dans la sous catégorie, il en est de même du troisième.

**Définition 6.1**. — Un module instable est localement fini si il est limite directe de modules finis.

**Proposition 6.2**. — *La sous-catégorie pleine des modules instables localement finis est épaisse.*

On notera que cette proposition est fausse sans l'hypothèse d'instabilité, comme le montre l'exemple suivant, où $\overline{\mathcal{A}}_2$ est l'idéal d'augmentation de $\mathcal{A}_2$ :

$$0 \longrightarrow \overline{\mathcal{A}}_2/\overline{\mathcal{A}}_2.\overline{\mathcal{A}}_2 \longrightarrow \mathcal{A}_2/\overline{\mathcal{A}}_2.\overline{\mathcal{A}}_2 \longrightarrow \mathbb{F}_2 \longrightarrow 0.$$

Un module instable $M$ est la suspension d'un module instable $N$ si, et seulement si, $\mathrm{Sq}_0 x = 0$ pour tout $x$ de $M$. En fait, pour un module instable $M$ quelconque, le sous-module $\{x \in M \mid \mathrm{Sq}_0 x = 0\}$ est la plus grande suspension contenue dans $M$. Plus généralement, les sous-objets suivants de $M$

$$M_k = \{x \in M \mid \mathrm{Sq}_0^k x = 0\}$$

forment une suite croissante de sous-modules instables de $M$, et chaque quotient successif est une suspension. Un module est dit nilpotent quand la suite des $M_k$ est convergente :

**Définition 6.3**. — Un module instable $M$ est nilpotent si, pour tout $x$ de $M$, il existe un entier $k$ tel que : $\mathrm{Sq}_0^k x = 0$ (resp. $\mathrm{P}_0^k x = 0$).

Une suspension est un module instable nilpotent ; en fait, la sous-catégorie des modules nilpotents est la plus petite sous-catégorie épaisse contenant toutes les suspensions et stable par limite directe.

**Proposition 6.4**. — *La sous-catégorie pleine des modules instables nilpotents est épaisse.*

C'est clair à partir de la seconde définition. On notera :

**Proposition 6.5**. — *Le produit tensoriel de deux modules nilpotents est nilpotent.*

**Définition 6.6**. — Un module instable est réduit si il ne contient pas de suspension non-triviale.

Par exemple, $\mathrm{H}^* \mathbb{Z}/p$ est réduit, et plus généralement $\mathrm{H}^* V$ l'est. Un module réduit ne contient aucun sous-module nilpotent non nul.

## 6.2. La catégorie $\mathcal{U}$ est localement noethérienne. 
— Le but de cette section est d'énoncer le théorème suivant, et de donner une brève indication de sa démonstration. On dit qu'un module sur l'algèbre de Steenrod est de type fini si, comme $\mathcal{A}_p$-module, il a un nombre fini de générateurs.

**Théorème 6.7**. — *Tout sous-module d'un module instable de type fini est encore de type fini.*

Un objet dans une catégorie abélienne est *noethérien* s'il a la propriété des chaînes croissantes. C'est-à-dire si toute suite croissante de sous-objets stabilise. Un ensemble de générateurs dans une catégorie abélienne $\mathcal{A}$ est un ensemble d'objets $\{M_\alpha\}$, $\alpha \in A$, tel que, pour tout objet $M$ de la catégorie, il y a un épimorphisme d'une somme directe des $M_\alpha$ sur $M$. Une catégorie est localement noethérienne si elle a un ensemble de générateurs noethériens [**9**]. Les modules instables $F(n)$ constituent un système de générateurs pour la catégorie $\mathcal{U}$. Donc, d'après le théorème 6.7, la catégorie $\mathcal{U}$ est localement noethérienne.

La démonstration repose sur une utilisation judicieuse des suites exactes :

$$0 \longrightarrow \mathrm{Sq}_0(F(n)) \longrightarrow F(n) \longrightarrow \Sigma F(n-1) \longrightarrow 0;$$

ces suites exactes résultent de la description de la base de Cartan-Serre. On démontre aussi, en utilisant les mêmes suites exactes :

**Proposition 6.8**. — *Le module instable $F(p) \otimes F(q)$ est de type fini pour tous $p$ et $q$. En conséquence, le produit tensoriel de deux modules instables de type fini l'est aussi.*

## 7. Représentabilité de foncteurs

Soit $\mathcal{C}$ une catégorie abélienne avec des produits. Un système de cogénérateurs dans $\mathcal{C}$ est un ensemble d'objets de $\mathcal{C}$, $\{C_\alpha, \alpha \in A\}$, tel que tout objet de $\mathcal{C}$ se plonge dans un produit de $C_\alpha$. Dans les sections 8 et 9, on introduit des cogénérateurs canoniques pour $\mathcal{U}$, et d'autres pour la sous-catégorie de $\mathcal{U}$ constituée par les modules réduits. Ce sont, par ailleurs, des objets injectifs dans $\mathcal{U}$. Rappelons la :

**Définition 7.1**. — Un objet $I$ d'une catégorie abélienne $\mathcal{C}$ est injectif si le foncteur

$$\mathcal{C} \longrightarrow \mathcal{A}b, \quad M \longmapsto \mathrm{Hom}_{\mathcal{C}}(M, I)$$

est exact.

Le produit au sens catégorique est défini par la propriété suivante :

$$\mathrm{Hom}_{\mathcal{C}}\left(M, \textstyle\prod_I B_i\right) \cong \prod_I \mathrm{Hom}_{\mathcal{C}}(M, B_i).$$

De la définition, il résulte qu'un produit quelconque d'objets injectifs est injectif. On a aussi :

**Proposition 7.2**. — *Toute limite directe filtrante, en particulier toute somme directe, d'objets injectifs dans une catégorie abélienne ayant un système de générateurs noethériens, est injective.*

Ce résultat s'applique à la catégorie $\mathcal{U}$. Les cogénérateurs que l'on considère sont de deux types :

– Les modules de Brown-Gitler, qui sont les cogénérateurs standards de la catégorie ;

– les modules de Carlsson, qui sont des cogénérateurs pour les modules instables réduits.

Ces modules sont construits comme objet classifiant de certains foncteurs de $\mathcal{U}^{\mathrm{op}}$ dans la catégorie **Vect** des $\mathbb{F}_p$-espaces vectoriels. Un foncteur $R$ est *représentable* s'il existe un module instable $B(R)$, appelé le *module classifiant* de $R$, et une équivalence naturelle de foncteurs $\gamma : \mathrm{Hom}_{\mathcal{U}}(-, B(R)) \to R$. Celle-ci se décrit comme suit : soit $v = \gamma(\mathrm{Id}_{B(R)})$ dans $R(B(R))$, à $f \in \mathrm{Hom}_{\mathcal{U}}(M, B(R))$, on associe $R(f)(v) \in R(M)$. Le lemme suivant donne une condition nécessaire et suffisante pour qu'un foncteur $\mathcal{U}^{\mathrm{op}}$ dans **Vect** soit représentable. C'est un cas particulier de résultats classiques [**9**].

**Lemme 7.3**. — *Le foncteur $R$ est représentable si, et seulement si, il est exact à droite et transforme les sommes directes en produits.*

Il est facile de voir que les conditions du lemme sont nécessaires. Considérons l'implication inverse et supposons le problème résolu : on a donc un module classifiant $B(R)$. Appliquons l'équivalence naturelle :

$$\gamma : R \longrightarrow \mathrm{Hom}_{\mathcal{U}}(-, B(R))$$

au module $F(n)$ ; on obtient :

$$R(F(n)) \cong \mathrm{Hom}_{\mathcal{U}}(F(n)), B(R)) = B(R)^n.$$

Ceci détermine $B(R)$ comme espace vectoriel gradué. La structure de module instable est décrite ainsi : pour $\theta$ dans $\mathcal{A}_p$ et $x$ dans $B(R)^n$,

$$\theta x = R(u_\theta)(x) \in B(R)^{n+|\theta|},$$

où $u_\theta : F(n + |\theta|) \to F(n)$ est l'application $\mathcal{A}_p$-linéaire associée à l'élément $\theta i_n \in F(n)^{n+|\theta|}$. Cette analyse suggère de définir une transformation naturelle de $R$ vers $\mathrm{Hom}_{\mathcal{U}}(-, B(R))$ par la formule suivante. Soit $M$ un objet dans $\mathcal{U}$ et $x$ dans $R(M)$. Un élément $y$ de $M$ peut être identifié avec une application $\widehat{y} : F(|y|) \to M$. On définit $\gamma_M(x)$ comme étant l'application

$$y \in M \longmapsto R(\widehat{y})(x) \in R(F(|y|) = B(R)^{|y|}.$$

Par construction, $\gamma_{F(n)}$ est un isomorphisme. Pour montrer que $\gamma_M$ est un isomorphisme, on considère le début d'une résolution projective de $M$ :

$$P_1 \longrightarrow P_0 \longrightarrow M \longrightarrow 0,$$

où $P_0$ et $P_1$ sont des sommes directes de $F(n)$. On a alors le diagramme commutatif :

$$
\begin{array}{ccccccc}
0 & \longrightarrow & R(M) & \longrightarrow & R(L_0) & \longrightarrow & R(L_1) \\
& & \gamma_M \downarrow & & \gamma_{L_0} \downarrow & & \gamma_{L_1} \downarrow \\
0 & \longrightarrow & \mathrm{Hom}_{\mathcal{U}}(M, B(R)) & \longrightarrow & \mathrm{Hom}_{\mathcal{U}}(L_0, B(R)) & \longrightarrow & \mathrm{Hom}_{\mathcal{U}}(L_1, B(R))
\end{array}
$$

La fin de la démonstration résulte alors du fait que $\gamma_{L_0}$ et $\gamma_{L_1}$ sont des isomorphismes, car $R$ transforme les sommes directes en produits, et d'une chasse au diagramme.

## 8. Modules de Brown-Gitler et l'algèbre de Miller

**8.1. Modules de Brown-Gitler.** — On considère maintenant des foncteurs satisfaisant aux hypothèses du lemme 7.3.

Le foncteur :

$$H_n : M \longmapsto \mathrm{Hom}_{\mathbf{Vect}}(M^n, \mathbb{F}_p)$$

est exact à droite et transforme les sommes directes en produits ; il est donc représentable. Comme il est exact à gauche, son module instable classifiant est injectif.

**Définition 8.1**. — Le $n$-ième module de Brown-Gitler $J(n)$ est le module instable classifiant pour le foncteur $H_n$.

Ce qui suit donne les propriétés de base des modules de Brown-Gitler.
- $J(0)$ est isomorphe à $\mathbb{F}_p$ ;
- $J(n)^m = F(m)^{n\#}$ ; en conséquence $J(n)$ est de dimension finie en chaque degré ;
- $J(n)^m = 0$ si $m > n$ ; c'est une conséquence de la propriété duale pour les $F(n)$ ;
- $J(n)^n \cong \mathbb{F}_p$ ; le générateur de $H_n(J(n))$ correspondant à l'application identité de $J(n)$ est noté $b_n$ ;
- l'application de restriction au degré $n$ :

$$\mathrm{Hom}_{\mathcal{U}}(M, J(n)) \longrightarrow \mathrm{Hom}_{\mathbf{Vect}}(M^n, J(n)^n)$$

s'identifie avec l'isomorphisme $\mathrm{Hom}_{\mathcal{U}}(M, J(n)) \cong (M^n)^\#$.

Cette remarque implique que, pour tout module instable $M$, l'application

$$M \longrightarrow \prod_n \prod_{u \in H_n(M)} J(n),$$

où la composante de l'application depuis $M$ vers le facteur indexé par $u$ du produit à droite est l'application déterminée par $u$, est injective. Donc la catégorie $\mathcal{U}$ a assez d'objets injectifs.

Une opération $\theta$ dans $\mathcal{A}_p$ détermine une application $\widetilde{\theta} : J(n+|\theta|) \to J(n)$ : il suffit de définir la forme linéaire associée, c'est

$$i_n \circ \theta : J(n+|\theta|)^n \longrightarrow \mathbb{F}_p,$$

où on rappelle que $\iota_n \in F(n)^n \cong (J(n)^n)^\#$ désigne le générateur canonique. Dans la suite on considère en particulier

$$\Lambda = \widetilde{\mathfrak{Sq}}^{n/2} : J(n) \quad \to J(n/2);$$

quand $n/2$ n'est pas entier, c'est par définition l'application nulle. On note aussi $\widetilde{\theta}$ l'application induite de $H_n(M)$ dans $H_{n+|\theta|}(M)$. On vérifie que :

**Proposition 8.2**. — *L'application $\widetilde{\theta} : H_n(M) \to H_{n+|\theta|}(M)$ est naturelle en $M$.*

La classe $\iota_m \otimes \iota_n \in H_{m+n}(J(m) \otimes J(n))$ détermine une application $\mathcal{A}_p$-linéaire

$$\mu_{m,n} : J(m) \otimes J(n) \longrightarrow J(m+n).$$

En degré $m+n$ l'application $\mu_{m,n}$ est l'identité de $\mathbb{F}_p$. Pour $m = 1$, comme $J(1) = \Sigma\mathbb{F}_p$, on obtient une application $\mu_{1,n} : \Sigma J(n) \to J(n+1)$.

**8.2. L'algèbre de Miller $J_*^*$.** — Pour comprendre la structure des modules de Brown-Gitler, il est commode, à la suite de H. Miller, d'introduire l'objet $\mathbb{N} \times \mathbb{N}$-bigradué, $J_*^*$, défini par $J_k^\ell = J(k)^\ell$, où le bidegré de $x \in J_k^\ell$ est noté $\|x\| = (\ell, k)$.

Cet objet a une structure très riche :

– c'est un module instable, car somme directe des modules $J(k)$, l'action de $\mathcal{A}_p$ laissant fixe le second degré ;

– il y a une application $\Lambda : J_*^* \to J_*^*$ (définie après la définition 8.1) qui laisse fixe le premier degré et divise le second par $p$ ; cette application est $\mathcal{A}_p$-linéaire ;

– les applications $\mu_{m,n} : J(m) \otimes J(n) \to J(m+n)$ définissent une structure de $\mathbb{F}_p$-algèbre bigraduée commutative, associative, unitaire ;

– la formule de Cartan a lieu pour ce produit et $\Lambda$ est multiplicative.

L'algèbre $J_*^*$ n'est pas instable car la seconde condition n'est pas vérifiée. En fait :

**Lemme 8.3**. — *Pour tout $x$ de $J_*^*$, on a :* $\mathrm{Sq}^0 x = (\Lambda x)^2$, *resp.* $\mathrm{P}^0 x = (\Lambda x)^p$.

Donnons la démonstration de cette formule pour $p = 2$. Étant donné un module instable $M$, soit $S^2(M)$ sa seconde puissance symétrique, *i.e.* le quotient de $M \otimes M$ par le sous-module des éléments de la forme $x \otimes y + y \otimes x$. Considérons alors le diagramme :

$$
\begin{array}{ccccc}
J(2k) \otimes J(2k) & \xrightarrow{p} & S^2(J(2k)) & \xrightarrow{\widetilde{\mu}_{2k,2k}} & J(4k) \\
{\scriptstyle \Lambda \otimes \Lambda}\downarrow & & {\scriptstyle S^2\Lambda}\downarrow & & {\scriptstyle \Lambda}\downarrow \\
J(k) \otimes J(k) & \xrightarrow{p} & S^2(J(k)) & \xrightarrow{\widetilde{\mu}_{k,k}} & J(2k).
\end{array}
$$

**Lemme 8.4**. — *Ce diagramme est commutatif.*

Il suffit de vérifier la commutativité du diagramme suivant puis de passer au quotient :

$$
\begin{array}{ccc}
J(2k) \otimes J(2k) & \xrightarrow{\mu_{2k,2k}} & J(4k) \\
{\scriptstyle \Lambda \otimes \Lambda}\downarrow & & \downarrow{\scriptstyle \Lambda} \\
J(k) \otimes J(k) & \xrightarrow{\mu_{k,k}} & J(2k).
\end{array}
$$

La commutativité de ce dernier se démontre par dualité : on le fait pour le diagramme suivant :

$$
\begin{array}{ccc}
F(2k)^{2k} & \longrightarrow & F(k)^k \otimes F(k)^k \\
{\scriptstyle \mathrm{Sq}^{2k}}\downarrow & & \downarrow{\scriptstyle \mathrm{Sq}^k \otimes \mathrm{Sq}^k} \\
F(2k)^{4k} & \longrightarrow & F(k)^{2k} \otimes F(k)^{2k}.
\end{array}
$$

Dans ce diagramme, l'application $\mathrm{Sq}^k$ envoie $\iota_k$ vers $\mathrm{Sq}^k \iota_k$. Pour interpréter ces applications $\mathrm{Sq}^k$, on se sert du :

**Lemme 8.5**. — *Soient $\theta$ dans $\mathcal{A}_p$, $M$ un module instable, et $u$ dans $H_n(M) = \mathrm{Hom}_{\mathcal{U}}(M, J(n))$. L'application $\widetilde{\theta} \circ u \in \mathrm{Hom}_{\mathcal{U}}(M, J(n - |\theta|)) \cong H_{n-|\theta|}(M)$ est la composée*

$$M^{n-|\theta|} \xrightarrow{\ \theta\ } M^n \xrightarrow{\ u\ } \mathbb{F}_p.$$

La démonstration résulte de la naturalité de $\widetilde{\theta}$. Pour achever la démonstration de la commutativité, on utilise l'instabilité et la formule de Cartan.

Voici la structure de $J_*^*$ :

**Théorème 8.6**. — *L'algèbre $J_*^*$ est isomorphe à l'algèbre polynomiale bigraduée*

$$\mathbb{F}_2[x_0, \ldots, x_i, \ldots]$$

*en des générateurs $x_i$ de bidegré $(1, 2^i)$. L'action de $\mathcal{A}_p$ est donnée par :*

- *$J_*^*$ est un module instable ;*
- *la formule de Cartan a lieu ;*
- *$\mathrm{Sq}^1(x_i) = x_{i-1}^2$ si $i > 0$ et $\mathrm{Sq}^1(x_0) = 0$ ;*
- *l'application d'algèbre $\Lambda$, déterminée par $\Lambda(x_0) = 0$ et $\Lambda(x_i) = x_{i-1}$, $i > 0$, est $\mathcal{A}_2$-linéaire.*

La démonstration est conséquence du théorème 3.1. Considérons le module bigra-dué $L_*^*$ défini par $L_j^i = F(j)^i$. À cause du théorème 3.1, on constate que $L_*^*$ s'iden-tifie avec l'algèbre à puissance divisée $\Gamma^*(F(1))$. En effet, par définition, $\Gamma^n(F(1)) \cong (F(1)^{\otimes n})^{\mathfrak{S}_n} \cong F(n)$. L'algèbre à puissances divisées est une algèbre de Hopf dont le produit le produit de battage (*shuffle produit*). On peut vérifier que le coproduit est, lui, donné par l'application canonique $F(p+q) \to F(p) \otimes F(q)$ déterminé par la seule classe non-triviale en degré $p + q$ de $F(p) \otimes F(q)$. En effet, l'application de coproduit

$$\Gamma^{p+q}(V) \longrightarrow \Gamma^p(V) \otimes \Gamma^q(V)$$

est donnée par la restriction des invariants sous $\mathfrak{S}_{p+q}$ aux invariants sous $\mathfrak{S}_p \times \mathfrak{S}_q$ :

$$(V^{\otimes(p+q)})^{\mathfrak{S}_{p+q}} \subset (V^{\otimes(p+q)})^{\mathfrak{S}_p \times \mathfrak{S}_q} \cong (V^{\otimes p})^{\mathfrak{S}_p} \otimes (V^{\otimes q})^{\mathfrak{S}_q}$$

Les applications $\widehat{\theta} : F(n + |\theta|) \to F(n)$ définies par les classes $\theta \iota_n$ déterminent une structure de module à droite sur l'algèbre de Steenrod sur $\Gamma^*(F(1))$, et cette action laisse fixe le degré supérieur.

Par dualisation on obtient l'objet $J_*^*$. Or, le dual de l'algèbre à puissances divisées $\Gamma^*(F(1))$ est l'algèbre symétrique $S^*(F(1)^\#)$, si bien que :

$$J_*^* \cong S^*(F(1)^\#).$$

Les générateurs $x_i$ sont donc les classes duales des générateurs de $F(1)$.

L'action à droite de $\mathcal{A}_2$ donne par dualisation l'action à gauche sur $J_*^*$. Enfin, on note que l'application $\Lambda$ est obtenue par dualisation, comme cas particulier, de l'action à gauche de $\mathcal{A}_2$ sur $\Gamma^*(F(1))$.

**Théorème 8.7** ([**13, 19**]). — *Soit $p > 2$, l'algèbre $J_*^*$ est isomorphe au quotient de la $\mathbb{F}_p$-algèbre commutative bigraduée libre engendrée par les éléments $e$, $x_i$, $i \geqslant 0$, $t_i$, $i \geqslant 0$, par les relations $e^2 = x_0$, où $\|e\| = (1,1)$, $\|x_i\| = (2, 2p^i)$, $\|t_i\| = (1, 2p^i)$. De plus :*

– *la structure de module instable à gauche est déterminée par $\mathrm{P}^1 x_i = x_{i-1}^p$ (avec $x_{-1} = 0$), $\mathrm{P}^1 t_i = 0$, $\mathrm{P}^1 e = 0$, $\beta t_i = x_i$, $\beta e = 0$ et la formule de Cartan ;*

– *l'application d'algèbre $\Lambda$ est déterminée par $\Lambda x_i = x_{i-1}$, $\Lambda t_i = t_{i-1}$, si $i > 0$, $\Lambda t_0 = e$, $\Lambda e = 0$, et est $\mathcal{A}_p$-linéaire.*

## 9. Modules de Carlsson

Supposons que $p = 2$ et considérons maintenant le foncteur :

$$K_i : M \longmapsto (\mathrm{colim}_q \{M^{2^q i}, \mathrm{Sq}^{2^q i}\})^\#.$$

On observe d'abord que l'on a des isomorphismes :

$$K_i(M) \cong \lim_q \{H_{2^q i}(M)^{2^q i}, \widetilde{\mathrm{Sq}}^{2^q i}\} \cong \lim_q \{H_{2^q i}(M)^{2^q i}, \Lambda\}.$$

Le foncteur est exact à droite et transforme les sommes directes en produits, il est donc représentable. Comme il est exact à gauche, le module classifiant est injectif.

**Définition 9.1**. — Le module de Carlsson $K(i)$ est le module classifiant pour le foncteur $K_i$.

On observe que :

**Proposition 9.2**. — *Le module instable $K(i)$ est isomorphe à la limite suivante :* $\lim_q \{J(2^q i), \Lambda\}$.

**Lemme 9.3**. — *En un degré donné, la limite précédente stabilise ; en particulier $K(i)$ est de dimension finie en chaque degré.*

Par construction on a $K(i) \cong K(2i)$. Ceci permet de définir $K(i)$ pour $i$ dans $\mathbb{N}[\frac{1}{2}]$. La signification de ces modules est donnée par :

**Proposition 9.4**. — *Un module $M$ est nilpotent si, et seulement si, pour tout entier $i$*

$$\mathrm{Hom}_{\mathcal{U}}(M, K(i)) = \{0\}.$$

On raisonne par contraposée. Pour $p = 2$, si $M$ n'est pas nilpotent, il existe $x$ dans $M^i$ tel que, pour tout $k$ : $\mathrm{Sq}_0^k x \neq 0$. Donc $K_i(M) \neq \{0\}$, et il existe une application non-triviale de $M$ dans $K(i)$.

On vérifie :

**Proposition 9.5**. — *Pour tout module instable réduit $M$, l'application*

$$M \longrightarrow \prod_{i \geqslant 0} \prod_{u \in K_i(M)} K(i),$$

*où la composante de l'application depuis $M$ vers le facteur indexé par $u$ du produit à droite est l'application déterminée par $u$, est injective.*

On introduit un objet $\mathbb{N} \times \mathbb{N}[\frac{1}{2}]$-bigradué $K_*^*$ par : $K_i^j = K(i)^j$. Pour déterminer sa structure, on interprète alors le théorème 8.6 de la manière suivante. On commence par définir un module $\mathbb{N} \times \mathbb{N}[\frac{1}{2}]$-bigradué $J[q]_*^*$, $q \geqslant 0$, par : $J[q]_k^\ell \cong J(2^q k)^\ell$, $J[q]_k^\ell \cong 0$ si $2^q k$ n'est pas un entier. Les applications $\Lambda$ déterminent une application d'algèbres $\Lambda : J[q]_*^* \to J[q-1]_*^*$. D'après le théorème 8.6, on a

$$J[q]_*^* \cong \mathbb{F}_2[x_{0,q}, \ldots, x_{i,q}, \ldots]$$

en des générateurs $x_{i,q}$ de bidegré $(1, 2^{i-q})$. L'application $\Lambda$ est donnée par : $\Lambda(x_{i,q}) = x_{i,q-1}$.

**Proposition 9.6 ([15])**. — *On a des isomorphismes :*

$$K_*^* \cong \lim_q \{J[q]_*^*, \Lambda\} \cong \mathbb{F}_2[\widehat{x}_i, i \in \mathbb{Z}],$$

$\widehat{x}_i$ *de bidegré $(1, 2^i)$. L'action de $\mathcal{A}_2$ est déterminée par la relation :*

$$\mathrm{Sq}^1(\widehat{x}_i) = \widehat{x}_{i-1}^2,$$

*et la formule de Cartan.*

Nous n'avons pas développé le cas $p > 2$, mais voici le résultat final :

**Proposition 9.7**. — *On a un isomorphisme :*

$$K_*^* \cong \mathbb{F}_p[\widehat{x}_i, i \in \mathbb{Z}] \otimes E(\widehat{t}_i, i \in \mathbb{Z}),$$

$\widehat{r}_i$ *de bidegré $(2, 2p^i)$, $\widehat{t}_i$ de bidegré $(1, 2p^i)$. L'action de $\mathcal{A}_p$ est déterminée par les relations :*

$$\mathrm{P}^1(\widehat{x}_i) = \widehat{x}_{i-1}^p, \quad \beta(\widehat{t}_i) = \widehat{x}_i,$$

*et la formule de Cartan.*

## 10. Injectivité de $\mathrm{H}^*\mathbb{Z}/p$ et le foncteur $T$

### 10.1. Injectivité de $\mathrm{H}^*\mathbb{Z}/p$. 
— Le théorème fondamental suivant est dû à H. Miller [19] ; il est implicite dans le travail de G. Carlsson [5] et a été généralisé par Lannes et Zarati [15].

**Théorème 10.1**. — *Soit $V$ un $p$-groupe abélien élémentaire. Le module instable $\mathrm{H}^*V$ est un objet injectif de la catégorie $\mathcal{U}$.*

On va d'abord démontrer le théorème pour $\dim V = 1$. Pour ce faire, on montre que $\widetilde{\mathrm{H}}^*\mathbb{Z}/p$ est facteur direct dans $K_1^* \cong K(1)$ pour $p = 2$, et dans $\oplus_{i=1}^{p-1} K_{2i}^*$ pour $p > 2$. Il y a une application

$$\gamma : \widetilde{\mathrm{H}}^*\mathbb{Z}/2 \longrightarrow K_1^*,$$

respectivement

$$\gamma : \widetilde{\mathrm{H}}^*\mathbb{Z}/p \longrightarrow \oplus_{i=1}^{p-1} K_{2i}^*.$$

Elle peut se construire de deux manières. La première, due à Carlsson, vient de la définition des $K(i)$. Pour $p = 2$, soit $\gamma : \widetilde{\mathrm{H}}^*\mathbb{Z}/2 \to K_1^*$ l'application correspondant au générateur de

$$\mathrm{colim}_q \{ \mathrm{H}^{2^q}\mathbb{Z}/2, \mathrm{Sq}^{2^q} \}^{\#} \cong \mathbb{F}_2.$$

Pour $p > 2$, $\gamma = \oplus_{i=1}^{p-1} \gamma_i$, où $\gamma_i$ est l'application correspondant au générateur de

$$\mathrm{colim}_q \{ \mathrm{H}^{2p^q i}\mathbb{Z}/p, \mathrm{P}^{p^q i} \}^{\#} \cong \mathbb{F}_p.$$

Cette application est décrite explicitement par H. Campbell et P. Selick [**4**]. Pour $p = 2$, on pose :

$$\gamma(u^n) = \left[ \left( \sum_{i \in \mathbb{Z}} \widehat{x}_i \right)^n \right]_1,$$

où la somme à droite est la somme (finie !) de tous les monômes dont le second degré est 1 dans la somme formelle infinie $(\sum_{i \in \mathbb{Z}} \widehat{x}_i)^n$. Pour $p > 2$, on pose :

$$\gamma_i(x^n) = \left[ \left( \sum_{i \in \mathbb{Z}} \widehat{x}_i \right)^n \right]_{2i},$$

$$\text{resp.} \quad \gamma_i(tx^n) = \left[ \left( \sum_{i \in \mathbb{Z}} \widehat{t}_i \right) \left( \sum_{i \in \mathbb{Z}} \widehat{x}_i \right)^n \right]_{2i-1},$$

où la somme à droite est la somme (finie !) de tous les monômes dont le second degré est $2i$ dans la somme formelle infinie $\left( \sum_{i \in \mathbb{Z}} \widehat{x}_i \right)^n$, respectivement $2i-1$ dans la somme formelle infinie $\left( \sum_{i \in \mathbb{Z}} \widehat{t}_i \right) \left( \sum_{i \in \mathbb{Z}} \widehat{x}_i \right)^n$.

Soit alors $g : K_*^* \to \mathrm{H}^*\mathbb{Z}/p$ l'unique application d'algèbres telle que : $g(\widehat{x}_j) = x$ et $g(\widehat{t}_j) = t$. Cette application est $\mathcal{A}_p$-linéaire.

On montre maintenant que $g \circ \gamma : \widetilde{\mathrm{H}}^*\mathbb{Z}/p \to \widetilde{\mathrm{H}}^*\mathbb{Z}/p$ est un isomorphisme. Pour cela on observe par calcul direct que $g \circ \gamma$ est un isomorphisme en degré inférieur ou égal à $p - 1$. Puis on a :

**Lemme 10.2**. — *Un endomorphisme $\mathcal{A}_p$-linéaire de $\widetilde{\mathrm{H}}^*\mathbb{Z}/p$, qui est un isomorphisme en degré inférieur ou égal à $p - 1$, est un isomorphisme.*

Voici des indications de démonstration dans le cas $p = 2$ ; on laisse au lecteur le cas $p > 2$ en exercice. Soit $u^n$ un élément dans $\widetilde{\mathrm{H}}^*\mathbb{Z}/2$. On montre que l'on peut trouver une opération $\tau$ dans $\mathcal{A}_2$ telle que : $\tau(u^n) = u^{2^q}$ pour un certain $q$. Or, si $f(u) = u$, on a : $f(u^{2^q}) = u^{2^q}$, et donc $f(\tau(u^n)) = f(u^{2^q}) = u^{2^q} \neq 0$ ; donc $f(\tau(u^n)) = u^n$ puisque $\mathrm{H}^n\mathbb{Z}/2 \cong \mathbb{F}_2$. Ceci achève la démonstration du théorème 10.1 en dimension 1.

**10.2. Le foncteur** $T$. — Soit $L$ un module instable de dimension finie en chaque degré. La proposition suivante est un avatar du théorème du foncteur adjoint de Freyd :

**Proposition 10.3**. — *Le foncteur $M \mapsto L \otimes M$, de $\mathcal{U}$ dans elle même, a un foncteur adjoint à gauche ; on le note $N \mapsto (N : L)_{\mathcal{U}}$.*

En d'autres termes il y a une équivalence naturelle :

$$\mathrm{Hom}_{\mathcal{U}}((N : L)_{\mathcal{U}}, M) \cong \mathrm{Hom}_{\mathcal{U}}(N, L \otimes M)$$

pour tous les modules instables $M$ et $N$.

La condition à vérifier est la commutation du foncteur $M \mapsto L \otimes M$ aux limites. Elle résulte de ce que $L$ est de dimension finie en chaque degré. Le foncteur $N \mapsto (N : L)_{\mathcal{U}}$ doit être interprété comme un foncteur de division. La division par $\mathbb{F}_p$ est l'identité. Si $L = \mathrm{H}^* V$, J. Lannes a introduit ce foncteur et l'a noté $T_V$. Si $V = \mathbb{F}_p$ il est noté simplement $T$. Ce foncteur est central dans les travaux de Lannes sur la conjecture de Sullivan et dans ses conséquences. Voici sa première propriété fondamentale :

**Théorème 10.4** (**[13]**). — *Le foncteur $T$ is exact.*

Comme $\mathrm{H}^* \mathbb{Z}/p$ est isomorphe à $\mathbb{F}_p \oplus \widetilde{\mathrm{H}}^* \mathbb{Z}/p$, le foncteur $T$ est naturellement équivalent à la somme directe du foncteur identité et du foncteur division par $\widetilde{\mathrm{H}}^* \mathbb{Z}/p$. Ce dernier foncteur sera noté $\overline{T}$. Voici un autre résultat sur $T$ :

**Proposition 10.5**. — *L'application naturelle $T \Sigma M \to \Sigma T M$ est un isomorphisme.*

La seconde propriété fondamentale de $T$ est le fait qu'il commute aux produits tensoriels.

Dans toute la suite, on n'aura besoin que de la composante de $T$ de degré 0, que l'on note $t : t(M) := T^0(M)$. On ne va donc pas donner des indications de démonstration pour $T$, mais seulement pour $t$. Il vaut cependant de noter que l'économie ainsi réalisée n'est pas uniforme selon les propriétés à démontrer. Le cas général dépend du fait que $\mathrm{H}^* \mathbb{Z}/p \otimes J(n)$, et plus généralement $\mathrm{H}^* E \otimes J(n)$, est un module instable injectif [**15**].

On a donc :

$$\mathrm{Hom}_{\mathbf{Vect}}(t(M), \mathbb{F}_p) \cong \mathrm{Hom}_{\mathcal{U}}(M, \mathrm{H}^* \mathbb{Z}/p),$$

et, plus généralement :

$$\mathrm{Hom}_{\mathbf{Vect}}(t(M), N^0) \cong \mathrm{Hom}_{\mathcal{U}}(M, \mathrm{H}^* \mathbb{Z}/p \otimes N^0),$$

pour tout module instable $N^0$ concentré en degré 0. On aura besoin de :

**Théorème 10.6** (**[10]**). — *Le foncteur $t : \mathcal{U} \to \mathbf{Vect}$, $M \mapsto t(M)$ est exact.*

Soit

$$0 \longrightarrow M' \longrightarrow M \longrightarrow M'' \longrightarrow 0$$

une suite exacte de modules instables. Dire que

$$0 \longrightarrow t(M') \longrightarrow t(M) \longrightarrow t(M'') \longrightarrow 0$$

est exacte équivaut à dire que :

$$0 \longrightarrow \mathrm{Hom}_{\mathcal{U}}(M'', \mathrm{H}^*\mathbb{Z}/p) \longrightarrow \mathrm{Hom}_{\mathcal{U}}(M, \mathrm{H}^*\mathbb{Z}/p) \longrightarrow \mathrm{Hom}_{\mathcal{U}}(M', \mathrm{H}^*\mathbb{Z}/p) \longrightarrow 0$$

est exacte. Ceci est équivalent au théorème 10.1 en dimension 1.

## 10.3. Exemples de calculs avec $t$

**Proposition 10.7**. — *Pour tout $n$, $t(F(n))$ est isomorphe à $\mathbb{F}_p$.*

Il suffit d'appliquer la définition de $F(n)$. En effet, $\mathrm{Hom}_{\mathcal{U}}(T(F(n)), \mathbb{F}_p)$ est isomorphe à

$$\mathrm{Hom}_{\mathcal{U}}(F(n), \mathrm{H}^*\mathbb{Z}/p),$$

qui est lui-même isomorphe à $\mathbb{F}_p$.

**Proposition 10.8**. — *Le foncteur $t$ commute aux colimites.*

Cela a lieu pour tous les foncteurs adjoints à gauche. Également :

**Théorème 10.9**. — *Si $M$ est localement fini, alors $t(M)$ est nul.*

En effet, le sous-module engendré sous $\mathcal{A}_p$ par une classe non-nulle dans $\widetilde{\mathrm{H}}^*\mathbb{Z}/p$ est infini. Donc, il ne peut y avoir d'application non-nulle d'un module instable fini dans $\widetilde{\mathrm{H}}^*\mathbb{Z}/p$. Le résultat suit, grâce à la décomposition $\mathrm{H}^*\mathbb{Z}/p \cong \widetilde{\mathrm{H}}^*\mathbb{Z}/p \oplus \mathbb{F}_p$ et par un argument de colimite.

## 10.4. Le foncteur $t$ et les produits tensoriels. — Le but de cette section est de démontrer le :

**Théorème 10.10**. — *Pour tous modules instables $M$ et $N$, il y a un isomorphisme naturel*

$$t(M \otimes N) \cong t(M) \otimes t(N).$$

Il faut d'abord définir une application naturelle

$$t(M \otimes N) \longrightarrow t(M) \otimes t(N).$$

On procède comme suit. On considère les adjoints $M \to t(M) \otimes \mathrm{H}^*\mathbb{Z}/p$, resp. $N \to t(N) \otimes \mathrm{H}^*\mathbb{Z}/p$, de l'identité de $t(M)$, resp. de $t(N)$, on en fait le produit tensoriel :

$$M \otimes N \longrightarrow t(M) \otimes \mathrm{H}^*\mathbb{Z}/p \otimes t(N) \otimes \mathrm{H}^*\mathbb{Z}/p \xrightarrow{\mathrm{Id} \otimes \tau \otimes \mathrm{Id}} t(M) \otimes t(N) \otimes \mathrm{H}^*\mathbb{Z}/p \otimes \mathrm{H}^*\mathbb{Z}/p,$$

puis on compose par le produit $\mathrm{H}^*\mathbb{Z}/p \otimes \mathrm{H}^*\mathbb{Z}/p \to \mathrm{H}^*\mathbb{Z}/p$. On obtient donc

$$M \otimes N \longrightarrow t(M) \otimes t(N) \otimes \mathrm{H}^*\mathbb{Z}/p,$$

et donc

$$t(M \otimes N) \longrightarrow t(M) \otimes t(N).$$

L'étape suivante de la démonstration consiste à calculer cette application dans le cas où $M = F(p)$ et $N = F(q)$. Dans la mesure où, par définition des modules $F(n)$ et du foncteur $t$, $t(F(n)) \cong \mathbb{F}_2$, ce dernier calcul se réduit pour l'essentiel à celui de $\mathrm{Hom}_{\mathcal{U}}(F(p) \otimes F(q), \mathrm{H}^*\mathbb{Z}/p)$. Voici le lemme fondamental :

**Lemme 10.11**. — *On a un isomorphisme :* $\mathrm{Hom}_{\mathcal{U}}(F(1)^{\otimes n}, \mathrm{H}^*\mathbb{Z}/p) \cong \mathbb{F}_p$.

On va donner les étapes de la démonstration quand $p = 2$.

**Lemme 10.12**. — *On a une suite exacte*

$$0 \longrightarrow \mathcal{A}_2 \left( u \otimes \cdots \otimes u^{2^{n-1}} \right) \longrightarrow F(1)^{\otimes n} \longrightarrow N \longrightarrow 0,$$

*avec $N$ nilpotent.*

On ne donne pas la démonstration en détails, mais seulement des indications. Il suffit de montrer que, pour toute classe $u^{2^{a_1}} \otimes u^{2^{a_2}} \otimes \cdots \otimes u^{2^{a_n}}$, il existe une opération $\theta$ et un entier $k$ tels que : $\theta(u \otimes u^2 \otimes \cdots \otimes u^{2^{n-1}}) = \mathrm{Sq}_0^k(u^{2^{a_1}} \otimes u^{2^{a_2}} \otimes \cdots \otimes u^{2^{a_n}})$. L'opération $\theta$ est obtenue par composition d'opérations du type $P_t^s$ [**17**]. Le fait que, dans $u \otimes u^2 \otimes \cdots \otimes u^{2^{n-1}}$, les puissances de 2 soient deux à deux distinctes, est crucial.

Il n'y a pas d'homomorphisme non-nul d'un module nilpotent dans $\mathrm{H}^*\mathbb{Z}/2$, car ce dernier module est réduit et l'application $\mathrm{Sq}_0$ y est injective. On en déduit que :

$$\mathrm{Hom}_{\mathcal{U}} \left( \mathcal{A}_2 \left( u \otimes \cdots \otimes u^{2^{n-1}} \right), \mathrm{H}^*\mathbb{Z}/2 \right) \cong \mathrm{Hom}_{\mathcal{U}}(F(1)^{\otimes n}, \mathrm{H}^*\mathbb{Z}/2).$$

Or le premier terme est isomorphe à $\{0\}$ ou $\mathbb{F}_2$, le module source étant cyclique, et le module but de dimension 1 en chaque degré. Comme il y a une application non-triviale obtenue par produit, on obtient :

$$\mathrm{Hom}_{\mathcal{U}}(F(1)^{\otimes n}, \mathrm{H}^*\mathbb{Z}/2) \cong \mathbb{F}_2.$$

Comme $F(p) \otimes F(q) \subset F(1)^{\otimes(p+q)}$, le calcul de $\mathrm{Hom}_{\mathcal{U}}(F(p) \otimes F(q), \mathrm{H}^*\mathbb{Z}/p)$ suit, en utilisant l'existence d'une application non-triviale et l'injectivité de $\mathrm{H}^*\mathbb{Z}/2$ :

$$\mathrm{Hom}_{\mathcal{U}}(F(p) \otimes F(q), \mathrm{H}^*\mathbb{Z}/2) \cong \mathbb{F}_2.$$

## 11. Injectivité de $\mathrm{H}^*E$, la formule de Campbell et Selick, et la caractérisation des modules nilpotents

### 11.1. Homomorphisme de Frobenius et action tordue des opérations de Steenrod. — Cette section présente la démonstration de Campbell et Selick de l'injectivité du module instable $\mathrm{H}^*E$. La formule pour $\gamma$ donnée en 10.1 est un cas particulier de cette démonstration. On travaillera en $p = 2$. Soit $d$ la dimension de $E$, et soit $M_d$ le quotient de l'algèbre $K_*^*$ par l'idéal engendré par les relations $\widehat{x}_{i+d} = \widehat{x}_i$, pour tout $i$ de $\mathbb{Z}$. Cet idéal est stable sous l'action de l'algèbre de Steenrod. En tant qu'algèbre, $M_d$ est isomorphe à $\mathbb{F}_2[x_0, \ldots, x_{d-1}]$, où l'on note $x_{i'}$ la classe de $\widehat{x}_i$, $i'$ étant le reste de la division de $i$ par $d$. Comme l'idéal par lequel on quotiente est stable sous l'algèbre de Steenrod, $M_d$ a une structure de module instable et la formule de

Cartan est satisfaite. Celle ci, avec les relations $\mathrm{Sq}^1 x_i = x_{i-1}^2$ (avec $\mathrm{Sq}^1 x_{d-1} = x_0^2$), détermine complètement la structure de module instable de $M_d$. On notera que $M_d$ n'est pas une algèbre instable.

Le résultat suivant est un remarquable théorème de rigidité concernant la structure du module instable de $\mathrm{H}^* E \cong \mathrm{H}^*(\mathbb{Z}/2^{\oplus d})$ :

**Théorème 11.1 ([4]).** — *Le module instable $M_d$ est isomorphe à $\mathrm{H}^*(\mathbb{Z}/2^{\oplus d})$.*

La démonstration est basée sur une extension des scalaires au corps fini $\mathbb{F}_{2^d}$. Soit $\omega$ un générateur du groupe multiplicatif de ce corps fini. La famille $(1, \omega, \dots, \omega^{d-1})$ en est une base sur $\mathbb{F}_2$. Considérons l'algèbre graduée $\mathbb{F}_{2^d} \otimes \mathrm{H}^* E \cong \mathbb{F}_{2^d}[t_0, \dots, t_{d-1}]$ $(\dim(E) = d)$. On y définit une action de $\mathcal{A}_2$ à partir de l'action standard sur $\mathrm{H}^* E \cong \mathbb{F}_2[t_0, \dots, t_{d-1}]$, en demandant que l'action de toute opération soit $\mathbb{F}_{2^d}$-linéaire. L'algèbre obtenue ainsi n'est pas instable. Notons $F$ l'homomorphisme de Frobenius agissant sur les coefficients des polynômes de $\mathbb{F}_{2^d}[t_0, \dots, t_{d-1}]$. Pour tout élément de degré 1, on a la relation

$$\mathrm{Sq}^1(Fx) = x^2.$$

Soit $T$ la matrice de la multiplication par $\omega$ dans la base $(1, \omega, \dots, \omega^{d-1})$ de $\mathbb{F}_{2^d}$, cette matrice est d'ordre $2^d - 1$ dans $\mathrm{GL}_d(\mathbb{F}_2)$. Considérons alors l'endomorphisme de $\mathbb{F}_{2^d}[t_0, \dots, t_{d-1}]^1$ (on considère les éléments de degré 1 de l'algèbre) dont la matrice dans la base $(t_0, \dots, t_{d-1})$ est $T$, et notons cet endomorphisme $\Theta$. Les valeurs propres de $\Theta$ sont $\omega, \omega^2, \dots, \omega^{2^{d-1}}$. Soit $x_0$ un vecteur propre associé à $\omega$. Si $F$ désigne comme plus haut l'homomorphisme de Frobenius, on constate que $F^j x_0$ est vecteur propre associé à $\omega^{2^j}$ ; on pose $x_j = F^j x_0$. On en déduit que $(x_0, \dots, x_{d-1})$ est une base de l'algèbre polynomiale graduée $\mathbb{F}_{2^d}[t_0, \dots, t_{d-1}]$. On a donc :

$$\mathbb{F}_{2^d}[t_0, \dots, t_{d-1}] = \mathbb{F}_{2^d}[x_0, \dots, x_{d-1}].$$

Le calcul montre alors que

$$\mathrm{Sq}^1 x_j = x_{j-1}^2.$$

On définit alors deux applications.

– D'abord l'application

$$\lambda : \mathbb{F}_2[x_0, \dots, x_{d-1}] \longrightarrow \mathbb{F}_2[t_0, \dots, t_{d-1}],$$

par $\lambda(x) = \zeta x + F(\zeta x) + \cdots + F^{d-1}(\zeta x)$, pour $\zeta$ dans $\mathbb{F}_{2^d}$, où on a choisi $\zeta$ pour que la famille $(\zeta, \zeta^2, \dots, \zeta^{2^{d-1}})$ soit une base de $\mathbb{F}_{2^d}$ sur $\mathbb{F}_2$.

– Puis l'application $\psi : \mathbb{F}_2[t_0, \dots, t_{d-1}] \to \mathbb{F}_2[x_0, \dots, x_{d-1}]$, définie comme suit : tout $x$ dans $\mathbb{F}_{2^d}[t_0, \dots, t_{d-1}]$ s'écrit de manière unique $x = \zeta q_0 + \cdots + \zeta^{2^{d-1}} q_{d-1}$, avec $q_i \in \mathbb{F}_2[x_0, \dots, x_{d-1}]$ ; alors pour $x$ dans $\mathbb{F}_2[t_0, \dots, t_{d-1}]$, on pose : $\psi(x) = q_0$.

Du fait que $x$ est dans $\mathbb{F}_2[t_0, \dots, t_{d-1}]$, on déduit que $q_i = F^i(q_0)$. Par construction, il en découle que $\lambda \circ \psi$ est l'identité, et le résultat suit.

**11.2. Le scindement des modules $K(i)$.** — Voici le second résultat de Campbell et Selick. Cet énoncé donne l'injectivité de $\mathrm{H}^*E$ et restitue le résultat central de [**14**], qui affirme que la famille des $\mathrm{H}^*E$ est équivalente à la famille des $K(i)$.

On définit sur l'algèbre $M_d$ une seconde graduation, à valeurs dans $\mathbb{Z}/2^d - 1$. Ce second degré est défini comme suit : on pose $w(x_i) = 2^i - 1 \in \mathbb{Z}/2^d - 1$, et on étend par produit. Le sous-espace $M_d(j) \subset M(d)$ engendré par les monômes de second degré égal à $j$ est stable par l'algèbre de Steenrod. On a :

$$M_d \cong \sum_{j \in \mathbb{Z}/2^d - 1} M_d(j).$$

Dans le théorème suivant, $j$ désigne soit une classe de congruence modulo $2^d - 1$, soit le représentant compris entre $0$ et $2^d - 1$ ; le contexte rend clair la bonne interprétation.

**Théorème 11.2 ([5]).** — *Le module instable $M_d$ est facteur direct dans $K_*^*$. Plus précisément, soit $\varphi_j$ l'application $M_d(j) \to K_j^* \cong K(j)$ qui, à $p(x_0, \ldots, x_{d-1})$, associe $\left[p\left(\sum_{i \equiv 0\,(d)} \widehat{x}_i, \ldots, \sum_{i \equiv d-1\,(d)} \widehat{x}_i\right)\right]_j$, la sous-somme (finie !) de la somme formelle infinie*

$$p\left(\sum_{i \equiv 0\,(d)} \widehat{x}_i, \ldots, \sum_{i \equiv d-1\,(d)} \widehat{x}_i\right),$$

*constituée par les monômes dont le second degré est égal à $j$ ; et soit $\gamma_j$ la restriction à $K_j^*$ de l'application d'algèbres, de $K_*^*$ dans $M_d$, qui à $\widehat{x}_i$ associe $x_{i'}$, où $i'$ est le reste de $i$ dans la division par $d$. Alors $\gamma_j \circ \varphi_j$ est l'identité de $M_d(j)$.*

Un moment de réflexion montre que si on calcule $\gamma_j \circ \varphi_j$ sur un monôme $m$, on a nécessairement : $\gamma_j \circ \varphi_j(m) = \varepsilon_m m$. Il faut montrer que $\varepsilon_m = 1$. Pour faire cela, Campbell et Selick se ramènent au cas où le monôme est produit de variables deux à deux distinctes à la puissance 1. L'argument de base est le lemme suivant :

**Lemme 11.3.** — *Soit $\tau$ l'application linéaire définie par :*

$$\tau : M_d \longrightarrow M_d, \quad x_{i_1}^{2^{r_1}} \cdots x_{i_k}^{2^{r_k}} \longmapsto x_{i_1 + r_1} \cdots x_{i_k + r_k}.$$

*Alors, pour tout monôme $m$, on a : $\varepsilon_{\tau(m)} = \varepsilon_m$.*

Par application itérée de ce lemme on se ramène au cas décrit plus haut. Dans ce cas on constate facilement que $\varphi_j(m)$ sur un monôme de second degré $j$ est constitué d'un seul monôme.

**11.3. Injectivité de $\mathrm{H}^*E$ et modules nilpotents, le foncteur $t_E$.** — Le module $M_d$, facteur direct dans un module injectif, est injectif. D'après l'énoncé 11.1, il est isomorphe, en tant que module instable, à $\mathrm{H}^*E$ ($\dim(E) = d$). Donc :

**Corollaire 11.4 ([15]).** — *Le module instable $\mathrm{H}^*E \cong \mathrm{H}^*\mathbb{Z}/2^{\otimes d}$ est injectif dans la catégorie $\mathcal{U}$.*

***Corollaire 11.5***. — *Le module $K_*^*$ est limite directe des modules $\mathrm{H}^*E$.*

On déduit alors de la proposition 9.4 :

***Théorème 11.6 ([14])***. — *Un module instable est nilpotent si, et seulement si, pour tout 2-groupe abélien élémentaire $E$ on a :*

$$\mathrm{Hom}_{\mathcal{U}}(M, \mathrm{H}^*E) = \{0\}.$$

Ce théorème a évidemment lieu pour $p > 2$. Définissons un foncteur $t_E$ par l'adjonction :

$$\mathrm{Hom}_{\mathbf{Vect}}(t_E(M), N^0) = \mathrm{Hom}_{\mathcal{U}}(M, \mathrm{H}^*E \otimes N^0),$$

où $N^0$ est un module instable concentré en degré 0. On a donc :

$$t_E(M) = \mathrm{Hom}_{\mathcal{U}}(M, \mathrm{H}^*E)',$$

où $\mathrm{Hom}_{\mathcal{U}}(M, \mathrm{H}^*E)'$ désigne le dual topologique du $\mathbb{F}_p$-espace vectoriel profini

$$\mathrm{Hom}_{\mathcal{U}}(M, \mathrm{H}^*E) \cong \lim_{i \in I} \mathrm{Hom}_{\mathcal{U}}(M_i, \mathrm{H}^*E),$$

où on prend la limite sur les sous-modules instables de $M$ de type fini. Sous cette denière hypothèse, le $\mathbb{F}_p$-espace vectoriel $\mathrm{Hom}_{\mathcal{U}}(M_i, \mathrm{H}^*E)$ est de dimension finie.

***Théorème 11.7***. — *Le foncteur $t_E$ est exact.*

Cela résulte du corollaire 11.4. L'observation élémentaire suivante est fondamentale pour la suite :

***Proposition 11.8***. — *Une application linéaire $f : E \to E'$ induit une transformation naturelle $f_* : t_E \to t_{E'}$.*

***Proposition 11.9***. — *Le foncteur $t_E$ commute aux colimites et aux limites finies.*

On a donc, en appliquant le théorème 11.6 :

***Théorème 11.10***. — *Un module instable $M$ est nilpotent si, et seulement si, $t_E(M) = 0$ pour tout $E$.*

**11.4. $t_E$ et les produits tensoriels.** — Voici la seconde propriété fondamentale de $t_E$.

***Théorème 11.11***. — *Pour tous modules instables $M$ et $N$, il existe un isomorphisme naturel :*

$$t_E(M \otimes N) \cong t_E(M) \otimes t_E(N).$$

Il faut d'abord définir une application naturelle

$$t_E(M \otimes N) \longrightarrow t_E(M) \otimes t_E(N).$$

On part des applications identité : $t_E(M) \to t_E(M)$ et $t_E(N) \to t_E(N)$ ; on en prend les adjointes : $M \to t_E(M) \otimes \mathrm{H}^*E$ et $N \to t_E(N) \otimes \mathrm{H}^*E$. On prend le produit tensoriel, on échange les deux facteurs centraux du but, et on compose par le produit de $\mathrm{H}^*E$. On obtient en fin de compte une application :

$$M \otimes N \longrightarrow t_E(M) \otimes t_E(N) \otimes \mathrm{H}^*E,$$

dont on prend l'adjointe :

$$t_E(M \otimes N) \longrightarrow t_E(M) \otimes t_E(N).$$

Comme les modules $F(n)$ sont des générateurs pour la catégorie $\mathcal{U}$, on se convainc qu'il suffit, pour démontrer le théorème 11.11, de montrer qu'il y a isomorphisme pour $F(p)$ et $F(q)$. Ceci résulte des lemmes qui suivent. On rappelle d'abord :

**Lemme 11.12**. — *Le quotient de $F(1)^{\otimes n}$ par le sous-module instable engendré par la classe $u \otimes u^2 \otimes \cdots \otimes u^{2^{n-1}}$ est nilpotent.*

Puis :

**Lemme 11.13**. — *L'espace $t_E(F(1)^{\otimes n})$ est de dimension $\dim(E)^n$.*

Il n'y a pas d'application non nulle d'un module nilpotent dans $\mathrm{H}^*E$. Il en résulte que :

$$\mathrm{Hom}_{\mathcal{U}}(\mathcal{A}_2(u \otimes u^2 \otimes \cdots \otimes u^{2^{n-1}}), \mathrm{H}^*E) \cong \mathrm{Hom}_{\mathcal{U}}(F(1)^{\otimes n}, \mathrm{H}^*E).$$

Il reste à déterminer les valeurs possibles pour l'image de $u \otimes u^2 \otimes \cdots \otimes u^{2^{n-1}}$. On construit des applications :

$$F(1)^{\otimes n} \longrightarrow \mathrm{H}^*E \cong \mathrm{H}^*\mathbb{Z}/2^{\otimes d} \cong \mathbb{F}_2[y_1, \ldots, y_d]$$

comme suit. On plonge $F(1)^{\otimes n}$ dans $\mathrm{H}^*\mathbb{Z}/2^{\otimes n} \cong \mathbb{F}_2[x_1, \ldots, x_n]$, par produit tensoriel du plongement de $F(1)$ dans $\mathrm{H}^*\mathbb{Z}/2$. Puis, à une partition indexée $\tau = (S_1, \ldots, S_d)$ de l'ensemble $(1, \ldots, n)$ en $d$ sous-ensembles, on associe l'application

$$f_\tau : F(1)^{\otimes n} \longrightarrow \mathbb{F}_2[y_1, \ldots, y_d]$$

obtenue par composition du plongement précédent avec l'application d'algèbres

$$j_\tau : \mathbb{F}_2[x_1, \ldots, x_n] \longrightarrow \mathbb{F}_2[y_1, \ldots, y_d]$$

qui envoie $x_i$ sur $y_j$ si $i \in S_j$, $1 \leqslant i \leqslant n$, $1 \leqslant j \leqslant d$. On a donc :

$$f_\tau(u \otimes u^2 \otimes \cdots \otimes u^{2^{n-1}}) = j_\tau(x_1 \cdots x_n^{2^{n-1}}) = y_1^{\Sigma_{i \in S_1} 2^{a_i - 1}} \cdots y_d^{\Sigma_{i \in S_d} 2^{a_{ji} - 1}}.$$

Quand $\tau$ décrit l'ensemble des partitions indexées de l'ensemble $(1, \ldots, n)$ en $d$ sous-ensembles, ces applications engendrent un sous-espace de $t_E(F(1)^{\otimes n})$ de la dimension annoncée.

Il reste donc à montrer que l'on a ainsi décrit toutes les applications de $F(1)^{\otimes n}$ dans $\mathbb{F}_2[y_1, \ldots, y_d]$. Pour cela, il suffit de montrer que l'image, par une application $\mathcal{A}_2$-linéaire, de $u \otimes u^2 \otimes \cdots \otimes u^{2^{n-1}}$ dans $\mathbb{F}_2[y_1, \ldots, y_d]$ est somme de monômes du type considéré plus haut, c'est-à-dire de la forme :

$$y_1^{\Sigma_{i \in S_1} 2^{a_i - 1}} \cdots y_d^{\Sigma_{i \in S_d} 2^{a_i - 1}},$$

où $(S_1, \ldots, S_d)$ est une partition (indexée) de l'ensemble $(1, \ldots, n)$. La description ci-dessus équivaut à dire que chacune des puissances $2^i$, $0 \leqslant i \leqslant n-1$, apparaît une fois et seule parmi les exposants des $y_j$. En particulier, la somme des puissances de 2 qui apparaîssent en exposant de $y_j$ est la décomposition 2-adique de cet exposant.

Considérons une somme de monômes du type

$$y_1^{\Sigma_{i \in S_1} 2^{a_{ji} - 1}} \cdots y_d^{\Sigma_{i \in S_d} 2^{a_j - 1}}.$$

On suppose que, pour chaque variable, la somme des puissances de 2 en exposant de cette variable est la décomposition 2-adique de l'exposant ; on suppose aussi que la somme des exposants vaut $2^n - 1$. Pour que cette expression soit l'image par une application $\mathcal{A}_2$-linéaire de $u \otimes u^2 \otimes \cdots \otimes u^{2^{n-1}}$, il faut que les opérations de Steenrod qui annulent $u \otimes u^2 \otimes \cdots \otimes u^{2^{n-1}}$ l'annulent aussi.

Or, par exemple, le produit de deux opérations de Milnor $Q_i$ est nulle sur $u \otimes u^2 \otimes \cdots \otimes u^{2^{n-1}}$, mais est non nulle sur un monôme (et une somme de monômes) où la décomposition 2-adique de l'exposant de deux variables distinctes ferait apparaître l'entier 1 ; il ne peut apparaître que dans la décomposition 2-adique de l'exposant d'une seule variable. On continue le même raisonnement avec les opérations $P_t^s$. Ceci donne le résultat.

**Corollaire 11.14**. — *L'application naturelle $t_E(F(1)^{\otimes n}) \to t_E(F(1))^{\otimes n}$ est un isomorphisme.*

On a une application naturelle :

$$t_E(F(1)^{\otimes n}) \longrightarrow t_E(F(1))^{\otimes n}.$$

Le lemme précédent, et le fait que, par définition de $F(1)$, on ait $t_E(F(1)) \cong E$, montre que les deux espaces ont même dimension. Une vérification de routine montre que l'application est injective.

**Lemme 11.15**. — *On a une suite exacte :*

$$0 \longrightarrow F(p) \longrightarrow F(1)^{\otimes p} \xrightarrow{\;\oplus(\tau - 1)\;} \oplus_\tau F(1)^{\otimes p},$$

*où la somme est prise sur les transpositions $\tau$ dans $\mathfrak{S}_p$.*

Ce lemme est conséquence de l'isomorphisme $F(p) \cong (F(1)^{\otimes p})^{\mathfrak{S}_p}$. Le théorème 11.11 pour $M = F(p)$ et $N = M(q)$ est alors conséquence du lemme 11.15 et de diagrammes tels que :

$$0 \longrightarrow t_E(F(p) \otimes F(1)^{\otimes q}) \longrightarrow t_E(F(1)^{\otimes p} \otimes F(1)^{\otimes q}) \longrightarrow \bigoplus_\tau t_E(F(1)^{\otimes p} \otimes F(1)^{\otimes q})$$

$$0 \to t_E(F(p)) \otimes t_E(F(1)^{\otimes q}) \to t_E(F(1)^{\otimes p}) \otimes t_E(F(1)^{\otimes q})) \to \bigoplus_\tau t_E(F(1))^{\otimes p} \otimes t_E(F(1)^{\otimes q}).$$

Dans ce diagramme, les lignes sont exactes ; on peut donc appliquer le lemme des cinq.

**11.5. Le théorème de Lannes sur les algèbres instables.** — Soit $K$ une algèbre instable. Elle est limite directe de ses sous-algèbres, $K_\alpha$, engendrées, en tant qu'algèbres instables, par un nombre fini d'éléments ; les ensembles $\mathrm{Hom}_{\mathcal{K}}(K_\alpha, \mathrm{H}^*E)$ sont finis. Il en résulte que $\mathrm{Hom}_{\mathcal{K}}(K, \mathrm{H}^*E)$ est limite inverse d'ensembles finis et est donc muni d'une structure profinie.

Une $\mathbb{F}_p$-algèbre de Boole est une $\mathbb{F}_p$-algèbre commutative dans laquelle on a $x^p = x$ pour tout élément $x$. Une telle algèbre $B$ est isomorphe à l'ensemble des applications continues de son spectre $\mathrm{Hom}_{\mathbb{F}_p\text{-Alg}}(B, \mathbb{F}_p)$ dans le corps $\mathbb{F}_p$. Le spectre est muni d'une topologie profinie en le considérant comme limite inverse des spectres des sous-algèbres de $B$ engendrées par un nombre fini d'éléments. Le théorème dont il est question s'énonce comme suit :

**Théorème 11.16**. — *Soit $K$ une algèbre instable. L'espace vectoriel $t_E(K)$ est naturellement une algèbre de Boole, isomorphe à l'algèbre de Boole des fonctions continues de l'espace profini $\mathrm{Hom}_{\mathcal{K}}(K, \mathrm{H}^*E)$ dans le corps $\mathbb{F}_p$.*

Ce théorème est dû à Lannes [**13**] ; le cas particulier suivant est un cas particulier du théorème d'Adams-Gunawardena-Miller [**1**] et avait été établi indépendemment par Lannes et Zarati [**16**]. Rappelons d'abord (section 6) que $\mathrm{Hom}_{\mathcal{K}}(\mathrm{H}^*V, \mathrm{H}^*W) \cong \mathrm{Hom}(W, V)$. Ce théorème a aussi des relations avec [**2**].

**Théorème 11.17**. — *Soient $\mathrm{H}^*V$ et $\mathrm{H}^*W$ la cohomologie modulo 2 des 2-groupes abéliens élémentaires $V$ et $W$. L'extension linéaire canonique du plongement :*

$$\mathrm{Hom}(W, V) \cong \mathrm{Hom}_{\mathcal{K}}(\mathrm{H}^*V, \mathrm{H}^*W) \longrightarrow \mathrm{Hom}_{\mathcal{U}}(\mathrm{H}^*V, \mathrm{H}^*W)$$

*est un isomorphisme.*

Autrement dit on a un isomorphisme :

$$\mathbb{F}_p[\mathrm{Hom}(W, V)] \cong \mathrm{Hom}_{\mathcal{U}}(\mathrm{H}^*V, \mathrm{H}^*W).$$

Ce résultat se déduit du précédent en prenant $E = W$, $K = \mathrm{H}^*V$ et en dualisant. On notera que la topologie profinie est ici triviale !

L'essentiel de la démonstration du théorème 11.16 consiste à montrer que $t_E(K)$ est une algèbre de Boole. Une fois que cela est fait, on montre que son spectre s'identifie avec $\mathrm{Hom}_{\mathcal{K}}(K, \mathrm{H}^*E)$. L'espace vectoriel $t_E(K)$ a une structure d'algèbre induite par :

$$t_E(K) \otimes t_E(K) \longrightarrow t_E(K \otimes K) \longrightarrow t_E(K).$$

Il convient de vérifier la condition $x^p = x$ pour tout $x$ de $t_E(K)$. Pour simplifier les notations, on va supposer que $p = 2$.

Si $K$ est une algèbre instable de multiplication $\mu$, on note $\widetilde{\mu}$ l'application induite de $S^2(K)$ dans $K$ par passage au quotient. Soit $\Phi(K)$ la sous-algèbre de $K$ constituée par les éléments de la forme $x^2$. L'application qui envoie $x^2$ sur la classe de $x \otimes x$ dans $S^2(K)$ est bien définie, et elle est $\mathcal{A}_2$-linéaire. De plus, dans la mesure où $K/\Phi(K)$ est un module nilpotent, on constate facilement que $t_E(K) \cong t_E(\Phi(K))$. La propriété requise est alors conséquence du lemme :

**Lemme 11.18**. — *La composée :*

$$t_E(K) \cong t_E(\Phi(K)) \xrightarrow{\lambda} t_E(S^2(K)) \xrightarrow{\widetilde{\mu}} t_E(K)$$

*est l'identité.*

Ce lemme se démontre en dualisant et en revenant aux définitions.

## 12. La catégorie $\mathcal{U}/\mathcal{N}il$ et les foncteurs analytiques

**12.1. L'équivalence de catégorie**. — Soit $\mathcal{C}$ une catégorie abélienne, et soit $\mathcal{D}$ une sous-catégorie pleine telle que, pour toute suite exacte dans $\mathcal{C}$

$$0 \longrightarrow C' \longrightarrow C \longrightarrow C'' \longrightarrow 0,$$

si deux objets sont dans $\mathcal{D}$, il en est de même du troisième. La sous-catégorie est dite *épaisse* (on dit aussi classe de Serre). On peut, sous cette hypothèse, définir la catégorie quotient $\mathcal{C}/\mathcal{D}$. Ses objets sont ceux de $\mathcal{C}$. Si $S$ et $T$ sont dans $\mathcal{C}$, $\mathrm{Hom}_{\mathcal{C}/\mathcal{D}}(S, T)$ est par définition

$$\mathrm{limdir}_{(S', T')} \mathrm{Hom}_{\mathcal{C}}(S', T/T'),$$

où $(S', T')$ parcourt l'ensemble des sous-objets $S'$ de $S$, $T'$ de $T$, tels que $S/S' \in \mathcal{D}$ et $T' \in \mathcal{D}$. Par ce procédé, un morphisme de $\mathcal{C}$, $\varphi : S \to T$ tel que $\mathrm{Ker}(\varphi) \in \mathcal{D}$ (resp. $\mathrm{Coker}(\varphi) \in \mathcal{D}$) devient un monomorphisme (resp. un epimorphisme) dans $\mathcal{C}/\mathcal{D}$. Le foncteur oubli

$$\mathcal{C} \longrightarrow \mathcal{C}/\mathcal{D}$$

est exact [**9**].

Soit $\mathcal{N}il$ la sous-catégorie pleine de $\mathcal{U}$ des modules nilpotents. La catégorie $\mathcal{N}il$ est une classe de Serre dans $\mathcal{U}$. On peut donc former la catégorie quotient $\mathcal{U}/\mathcal{N}il$. La sous-catégorie $\mathcal{N}il$ est stable par colimite, et tout module instable contient un plus grand sous-module nilpotent. Il en résulte que le foncteur oubli

$$\mathcal{U} \longrightarrow \mathcal{U}/\mathcal{N}il$$

a un adjoint à droite, $s$. Le module instable $sr(M)$ est réduit, et même $\mathcal{N}il$-fermé [**9**], [**21**] : il y a une application naturelle :

$$M \longrightarrow sr(M).$$

On note $\ell = sr$, c'est le foncteur de *localisation loin de $\mathcal{N}il$*.

Soit $\mathcal{F}$ la catégorie des foncteurs de la catégorie $\mathcal{V}^f$ des $\mathbb{F}_p$-espaces vectoriels de dimension finie vers la catégorie **Vect** des $\mathbb{F}_p$-espaces vectoriels. Elle est abélienne : une suite exacte de foncteurs est une suite $0 \to F' \to F \to F'' \to 0$, telle que, pour tout $V$,

$$0 \longrightarrow F'(V) \longrightarrow F(V) \longrightarrow F''(V) \longrightarrow 0$$

est exacte.

Soit

$$f : \mathcal{U} \longrightarrow \mathcal{F}$$

le foncteur qui à $M \in \mathcal{U}$ associe le foncteur

$$f(M) : \mathcal{V} \longrightarrow \mathbf{Vect}, \quad V \longmapsto t_V(M).$$

**Théorème 12.1**. — *Le foncteur $f$ est exact.*

L'exactitude de $t_E$ pour tout $E$ est équivalente à celle de $f$.

La catégorie $\mathcal{F}$ a un produit tensoriel. Le produit tensoriel de deux foncteurs $F$ et $G$ est le foncteur $F \otimes G$ qui, à $V$, associe le produit tensoriel $F(V) \otimes G(V)$. Il résulte de 11.11 que :

**Théorème 12.2**. — *Pour tous modules instables $M$ et $N$, il y a une équivalence naturelle de foncteurs*

$$f(M) \otimes f(N) \cong f(M \otimes N).$$

Le foncteur $f$ commute aux colimites et aux limites finies. Voici des exemples :

– on a

$$f(F(1)^{\otimes n})(V) \cong V^{\otimes n};$$

– pour $p = 2$ : si $M$ est le module instable $F(n)$,

$$f(F(n))(V) \cong H_n(V) \cong \Gamma^n(V),$$

la $n$-ième puissance divisée sur $V$ ;

– pour $p > 2$ : si $M$ est le module instable $F(2n)$,

$$f(F(2n))(V) \cong H_n(V) \cong \oplus_{2i+j=n}\Gamma^i(V) \otimes \Lambda^j(V) \ ;$$

une formule analogue a lieu pour $F(2n+1)$ ;

– l'exemple suivant est une reformulation du théorème d'Adams-Gunawardena-Miller. Le foncteur $f(\mathrm{H}^*E)$ est le foncteur $I_E$ qui à $V$ associe $\mathbb{F}_2^{\mathrm{Hom}(V,E)}$, l'espace des fonctions de $\mathrm{Hom}(V,E)$ dans $\mathbb{F}_2$.

On décrit maintenant le noyau de $f$ et son image dans $\mathcal{F}$. Si $M$ est nilpotent, $t_E(M)^0$ est trivial pour tout $E$. Ceci permet de définir :

$$\widetilde{f} : \mathcal{U}/\mathcal{N}il \longrightarrow \mathcal{F}.$$

Le théorème suivant résulte du théorème 11.10.

**Théorème 12.3 ([10]).** — *Le foncteur $\widetilde{f} : \mathcal{U}/\mathcal{N}il \to \mathcal{F}$ est fidèle.*

Il nous reste à décrire l'image de $f$. Pour $F$ dans $\mathcal{F}$, définissons $\Delta F$ comme étant le foncteur qui à $V$ associe le noyau de $F(\pi) : F(V \oplus \mathbb{F}_2) \to F(V)$, où $\pi : V \oplus \mathbb{F}_2 \to V$ est la projection .

**Définition 12.4.** — Un foncteur $F$ est de degré d'Eilenberg-MacLane moins que $k$ (resp. exactement $k$) si $\Delta^{k+1}F$ est le foncteur nul (resp. si $\Delta^{k+1}F$ est nul et $\Delta^k F$ est non nul).

On comparera à [**7**, 3.1 et 3.2].

**Définition 12.5.** — Un foncteur de $\mathcal{F}$ est polynomial s'il est polynomial de degré inférieur ou égal à $k$ pour un certain entier $k$. Un foncteur de $\mathcal{F}$ est analytique si il est limite directe de ses sous-foncteurs polynomiaux.

Soit $\mathcal{F}_n$ la sous-catégorie pleine de $\mathcal{F}$ des foncteurs de degré inférieur ou égal à $n$, et $\mathcal{F}_\omega$ la sous-catégorie pleine des foncteurs analytiques. Ce sont des sous-categories abéliennes de $\mathcal{F}$. Elles sont stables par colimites. Les foncteurs $V \mapsto V^{\otimes n}$ et $V \mapsto \Gamma^n(V)$ sont polynomiaux de degré $n$.

**Théorème 12.6.** — *Pour tout module instable $M$ le foncteur $f(M)$ est analytique.*

Il suffit de le vérifier sur un ensemble de générateurs de la catégorie $\mathcal{U}$. Or, c'est vrai pour les modules instables $F(n)$, pour tout $n$.

**Corollaire 12.7.** — *Le foncteur $I_E : V \mapsto \mathbb{F}_p^{\mathrm{Hom}(V,E)}$ est analytique pour tout $E$.*

En effet $f(\mathrm{H}^*E) = I_E$. Le résultat principal de cette section est le suivant :

**Théorème 12.8 ([10]).** — *Le foncteur induit par $f$ de $\mathcal{U}/\mathcal{N}il$ vers $\mathcal{F}_\omega$ est une équivalence de catégories.*

Voici un corollaire immédiat :

**Corollaire 12.9.** — *Les foncteurs puissances divisées $\Gamma^n$, $n \geqslant 0$, sont des générateurs pour $\mathcal{F}_\omega$.*

C'est clair puisque les $F(n)$ sont des générateurs pour $\mathcal{U}$. En appliquant la dualité de Kuhn qui, à un foncteur $F$, associe le foncteur $DF$ donné par $DF(V) = F(V^\#)^\#$, on obtient :

**Corollaire 12.10** ([11]). — *Tout foncteur polynomial se plonge dans une somme directe de puissances symétriques.*

Le résultat n'a pas lieu si on suppose seulement le foncteur analytique.

Pour démontrer le théorème 12.8 il faut montrer que le foncteur est « surjectif » sur les objets et les morphismes. Montrons le sur les objets. D'abord un argument de colimite permet de se réduire à montrer que pour un foncteur polynomial $F$, prenant des valeurs de dimension finie, on peut trouver un module instable $M$ tel que $f(M) \cong F$. Pour ce faire on observe que l'on peut construire le début d'une résolution injective de $F$ comme suit :

$$0 \longrightarrow F \longrightarrow I_0 \xrightarrow{\rho} I_1,$$

où $I_0$ et $I_1$ sont des sommes directes de foncteurs $I_E$, que l'on peut supposer finies (voir le théorème 12.22 ou [**20**, 3.3]. Les foncteurs $I_E$ sont des objets injectifs tautologiques de la catégorie car :

**Proposition 12.11**. — *Soit $E$ dans $\mathcal{V}^f$. Le foncteur $I_E : V \mapsto \mathbb{F}_2^{\mathrm{Hom}(V,E)}$ représente le foncteur exact $F \mapsto F(E)^\#$, $\mathcal{F} \to \mathbf{Vect}$. Il est donc injectif.*

Si $I_0 \cong \oplus_\alpha I_{E_\alpha}$ et $I_1 \cong \oplus_\beta I_{E_\beta}$ alors

$$\mathrm{Hom}_{\mathcal{F}}(I_0, I_1) \cong \bigoplus_{\alpha,\beta} \mathbb{F}_p[\mathrm{Hom}(E_\beta, E_\alpha)].$$

À cause du théorème d'Adams-Gunawardena-Miller (et des hypothèses de finitude), cet espace s'identifie à

$$\bigoplus_{\alpha,\beta} \mathrm{Hom}_{\mathcal{U}}(\oplus_\beta \mathrm{H}^* E_\beta, \oplus_\alpha \mathrm{H}^* E_\alpha).$$

Soit alors $r : \oplus_\alpha \mathrm{H}^* E_\alpha \to \oplus_\beta \mathrm{H}^* E_\beta$ l'élément de cet espace correspondant à $\rho$. Par hypothèse, $f(r) = \rho$ ; puis par l'exactitude de $f$ on obtient que :

$$f(\mathrm{Ker}\, r) \cong \mathrm{Ker}\, f(r) = \mathrm{Ker}(\rho) \cong F.$$

## 12.2. L'adjoint de $f$

**Proposition 12.12**. — *Le foncteur $f$ a un adjoint à droite $m$, qui satisfait donc à un isomorphisme naturel :*

$$\mathrm{Hom}_{\mathcal{F}}(f(M), F) \cong \mathrm{Hom}_{\mathcal{U}}(M, m(F)).$$

Ceci se déduit du lemme 7.3 appliqué au foncteur $M \mapsto \mathrm{Hom}_{\mathcal{F}}(f(M), F)$. Le foncteur $m$ est composé du foncteur section $s$, de l'équivalence inverse à $\mathcal{U}/\mathcal{N}il \cong \mathcal{F}_\omega$, et du foncteur qui à un objet de $\mathcal{F}$ associe son plus grand quotient analytique. Par application de la définition à $M = F(n)$, le foncteur $m$ est donné par la formule suivante :

$$m(F) \cong \operatorname{Hom}_{\mathcal{F}}(\Gamma^*, F),$$

qui définit $m(F)$ comme espace vectoriel gradué : $m(F)^n = \operatorname{Hom}_{\mathcal{F}}(\Gamma^n, F)$. La structure de module sur l'algèbre de Steenrod est donnée comme suit. Soit $\theta$ dans $\mathcal{A}_2^r$ et soit $x$ dans $m(F)^n$. À l'opération $\theta$, on associe l'application $\widehat{\theta} : F(n+r) \to F(n)$ qui, à $\iota_{n+r}$, associe $\theta \iota_n$. On en déduit, par application de $f$, une transformation naturelle $\Gamma^{n+r} \to \Gamma^n$. Alors $\theta(x)$ est la transformation naturelle composée $x \circ \widehat{\theta}$.

Il résulte des généralités de [**9**] et [**15**] que l'unité de l'adjonction $M \to m(f(M))$ a les propriétés suivantes :

– c'est un isomorphisme dans $\mathcal{U}/\mathcal{N}il$ ;
– $m(f(M))$ est réduit ;
– $\operatorname{Ext}_{\mathcal{U}}^1(\Sigma N, m(f(M)))$ pout tout $N$ dans $\mathcal{U}$, le module instable $m(f(M))$ est $\mathcal{N}il$-fermé.

On note $\ell$ le foncteur $m \circ f$. On a :

**Théorème 12.13**. — *Étant donnés $M$ dans $\mathcal{U}$ et $F$ dans $\mathcal{F}_\omega$, il existe une suite spectrale du premier quadrant :*

$$\operatorname{Ext}_{\mathcal{U}}^u(M, \ell^v(m(F))) \Longrightarrow \operatorname{Ext}_{\mathcal{F}}^{u+v}(f(M), F).$$

Ceci se relie aux résultats de [**20**]. En faisant, par exemple pour $p = 2$, $M = F(k)$, $F = S^n$, et en utilisant le fait que $F(k)$ est projectif dans $\mathcal{U}$, on obtient :

$$\ell^v(m(S^n))^k \cong \operatorname{Hom}_{\mathcal{U}}(F(k), \ell^v(m(S^n))) \cong \operatorname{Ext}_{\mathcal{F}}^v(\Gamma^k, S^n).$$

Le dernier terme est calculé dans [**20**] pour $k = 1$ et en général dans [**6**].

## 12.3. Le lemme d'annulation de Kuhn.

— La proposition 12.11 montre que $I_E$ est un objet injectif naturel dans $\mathcal{F}$. Alors qu'il est difficile de montrer que le module instable $\operatorname{H}^*E$, qui correspond à $I_E$ *via* $f$, est injectif (corollaire 11.4), cette propriété est tautologique pour $I_E$. D'un autre côté, les modules $K(i)$ sont des modules injectifs naturels dans $\mathcal{U}$ ; leur image par l'équivalence de catégories $f$ est donc injective, mais ce n'est pas formel dans $\mathcal{F}$. En fait, on a un résultat un peu plus précis :

**Proposition 12.14** ([**12**]). — *Le foncteur $f(K(i))$ est équivalent à la limite directe sur $q$, notée $S_i$, des foncteurs puissances symétriques $S^{2^q i}$ (resp. $S^{2p^q i}$) sous l'homomorphisme de Frobenius. Il en résulte que ce foncteur $S_i$ est injectif dans $\mathcal{F}_\omega$.*

On va seulement montrer que la limite directe $S_i$ est injective, laissant l'identification en exercice (qui donne une autre preuve de la dite injectivité !). L'injectivité des foncteurs $S_i$ dans la catégorie $\mathcal{F}$ reste une question ouverte.

On veut montrer que, si on a une application injective $f : F \to G$ et une application $r : F \to S_i$, on peut l'étendre en une application $r' : G \to S_i$. On commence par observer que l'on peut se restreindre, par un argument de limite directe, à des foncteurs

$F$ et $G$ polynomiaux, prenant des valeurs de dimension finie. Puis on applique la dualité de Kuhn $(DF(V) = F(V^\#)^\#)$.

Supposons $p = 2$. Par dualité la limite directe :

$$\cdots \xrightarrow{F} S^{2^q i} \xrightarrow{F} S^{2^{q+1} i} \xrightarrow{F} \cdots$$

se transforme en la limite inverse :

$$\cdots \xrightarrow{V} \Gamma^{2^{q+1} i} \xrightarrow{V} \Gamma^{2^q i} \xrightarrow{V} \cdots$$

où $V$ est le Verschiebung de l'algèbre à puissances divisées.

**Lemme 12.15**. — *Pour $p = 2$, l'application $V$ est obtenue par application de $f$ à l'application composée suivante :*

$$F(2n) \longrightarrow \mathrm{Sq}_0(F(n)) \subset F(n)$$

*où la première application est l'unique application non-nulle de $F(2n)$ dans $\mathrm{Sq}_0(F(n))$.*

Si $F$ est polynomial et prend des valeurs de dimension finie on obtient, en utilisant la proposition 12.12 :

$$\mathrm{Hom}_{\mathcal{F}}(F, S_i) = \mathrm{limdir}_q \, \mathrm{Hom}_{\mathcal{F}}(\Gamma^{2^q i}, DF) = \mathrm{limdir}_q \, \mathrm{Hom}_q(F(2^q i), m(DF))$$

$$= \mathrm{limdir}_q \, m(DF)^{2^q i}.$$

On déduit de ce qui précède que, pour démontrer la proposition 12.14, il suffit de montrer :

**Proposition 12.16**. — *Étant donnée une application $f : \Gamma^n \to F$ et une application surjective $p : DG \to DF$, il existe une application $g : \Gamma^{2^r n} \to DG$ telle que $p \circ g = f \circ V^r$.*

Ceci résulte des deux lemmes suivants.

**Lemme 12.17** (**[8]**). — *Si $M$ un module instable de type fini réduit, alors l'application*

$$\mathrm{Sq}_0 : M^q \longrightarrow M^{2q}$$

*est un isomorphisme pour tout $q$ assez grand.*

On le montre d'abord pour les modules $F(n)$, et c'est alors un exercice de combinatoire. Puis on étend le résultat à un module de type fini, à l'aide d'une résolution projective, en utilisant le fait que la catégorie $\mathcal{U}$ est localement noethérienne.

On montre ensuite que, si $H$ est polynomial et prend des valeurs de dimension finie, conditions conservées par dualité, $m(H)$ est réduit et de type fini. Que le module instable $m(H)$ soit réduit pour tout $H$ découle de la définition de $m$. Montrer que $m(H)$ est de type fini nécessite plus de travail.

**Proposition 12.18**. — *Soit $M$ un module instable de type fini, alors*

$$\text{limdir}_q \ M^{2^q i} \cong \text{limdir}_q \ m(f(M))^{2^q i},$$

*en fait il y a isomorphisme pour tout $q$ assez grand.*

On va seulement donner les deux arguments principaux de la démonstration. D'abord, la proposition est vraie pour les modules $M$ qui sont des sous-modules de $F(n)$ tels que $m(f(F(n)) = F(n)$. Pour montrer cela, on utilise que ces sous-modules sont de type fini, ce qui entraîne que, sous l'hypothèse sur $M$, il existe $k$ tel que $\text{Sq}_0^k(F(n) \subset M$. Puis, dans une seconde étape, on passe au cas général en considérant un module de type fini comme un quotient d'une somme directe finie de $F(n)$.

Revenons à la démonstration de 12.17. On déduit des deux lemmes précédents que, si $DG \to DF$ est surjective, $m(DG)^{2^q i} \to m(DF)^{2^q i}$ l'est aussi pour un entier $q$ assez grand. On en déduit 12.17.

Cette discussion pour le cas $p = 2$ reste valable dans le cas $p > 2$, une fois faite l'observation suivante. Les limites directes prises sur les duaux des foncteurs $f(F(2p^q i))$ le long du morphisme de Frobenius, ou de sa généralisation $P_0$, sont isomorphes aux limites directes prises sur les puissances symétriques (voir 12.1), car l'homomorphisme de Frobenius est nul sur les puissances extérieures.

**12.4. Un lemme de finitude.** — Si $E = \mathbb{F}_p$, le foncteur $I_{\mathbb{F}_p}$ se décompose canoniquement en somme directe du foncteur constant $\mathbb{F}_p$ (vu comme le sous-ensemble des fonctions constantes de $V^{\#}$ dans $\mathbb{F}_2$) et d'un foncteur $\widetilde{I}$. Par définition, $\text{Hom}(F, \widetilde{I})$ est canoniquement isomorphe à $\Delta F(0)$. Plus généralement, $\text{Hom}_{\mathcal{F}}(F, \widetilde{I}^{\otimes k})$ est canoniquement isomorphe à $\Delta^k F(0)$.

**Théorème 12.19**. — *La catégorie $\mathcal{F}_\omega$ est localement noethérienne. En particulier, toute somme directe $\oplus_\alpha E_{W_\alpha}$ est injective.*

Un foncteur $F \in \mathcal{F}$, polynomial, tel que $F(\mathbb{F}_p^{\oplus d})$ soit fini pour tout $d$, a une filtration finie $0 \subset F_0 \subset F_1 \subset ... \subset F_l = F$ dont les quotients sont simples (c'est-à-dire des foncteurs qui n'ont pas de sous-foncteurs non-triviaux).

Considérons, en effet, pour un foncteur $F$ quelconque, la fonction définie par

$$P_F(d) = \dim_{\mathbb{F}_p} F(\mathbb{F}_p^{\oplus d}).$$

Il est facile de vérifier que $P_{\Delta F}(d) = P_F(d + 1) - P_F(d)$; on en déduit que si $F$ est polynomial, $P_F$ est un polynôme. Puis on remarque qu'un polynôme qui prend des valeurs entières positives ou nulles sur les entiers positifs ou nuls ne peut être somme d'un nombre arbitrairement grands de tels polynômes. De plus, tous les foncteurs simples sont polynomiaux. Ceci résulte, par exemple, du fait qu'un foncteur simple se plonge dans un $I_E$. Ceci implique le résultat annoncé.

Le socle d'un objet dans une catégorie abélienne est son plus grand sous-objet semi-simple (isomorphe à une somme directe d'objets simples). Le socle $\text{Soc}(F)$ d'un foncteur analytique $F$ a la propriété que tout sous-foncteur $G \subset F$ a une intersection non-triviale avec lui. On montre ce dernier point comme suit. Soit $G$ un sous-foncteur non-trivial de $F$; comme $G$ est analytique, il contient un sous-foncteur polynomial non-trivial $F$ tel que $\dim_{\mathbb{F}_p} F(\mathbb{F}_p^{\oplus d})$ est fini pour tout $d$. Ce foncteur contient un sous-foncteur simple non-trivial. Donc $G$ et $\mathrm{Soc}(F)$ ont une intersection non-nulle.

**Théorème 12.20**. — *Un foncteur analytique se plonge dans une somme directe $\oplus_\alpha I_{E_\alpha}$. Si son socle est une somme directe finie, il se plonge dans une somme directe finie de $I_E$.*

Il est naturel de demander si cette propriété s'étend à une résolution injective. Voici une réponse partielle :

**Théorème 12.21**. — *Soit $F$ dans $\mathcal{F}_k$. Supposons que $P_F(d)$ soit fini pour tout $d$. Alors, il existe une résolution injective de $F$ :*

$$0 \longrightarrow F \longrightarrow I_0 \longrightarrow I_1 \longrightarrow \cdots \longrightarrow I_k \longrightarrow \cdots$$

*telle que tous les foncteurs $I_k$ sont des sommes directes finies de $I_E$.*

Dans cet énoncé, on pourrait remplacer la condition sur les $I_k$ par celle, un peu plus naturelle, qu'ils soient somme directe finie d'objets injectifs indécomposables.

La démonstration de ce résultat est donnée dans [**20**, 3.3] dans le présent volume ; nous y renvoyons, ainsi qu'à [**21**] et [**8**].

## 13. Compléments

**13.1. Le foncteur $p_n : \mathcal{U}/\mathcal{N}il \to \mathcal{M}od_{\mathbb{F}_p[\mathfrak{S}_n]}$, la filtration sur $\mathcal{U}/\mathcal{N}il$**. — On va pousser plus loin la comparaison des catégories $\mathcal{U}/\mathcal{N}il$ et $\mathcal{F}_\omega$. On suppose $p = 2$. On définit une filtration $\mathcal{U}/\mathcal{N}il$, et on l'identifie à la filtration polynomiale sur $\mathcal{F}_\omega$. Puis on identifie les quotients avec les catégories de représentations modulaires des groupes symétriques. Finalement, on décrit les objets simples dans $\mathcal{U}/\mathcal{N}il$ et dans $\mathcal{F}_\omega$.

Soit $\mathcal{M}od_{\mathbb{F}_2[\mathfrak{S}_n]}$ la catégorie odes $\mathbb{F}_2[\mathfrak{S}_n]$-modules à droite ($\mathfrak{S}_n$ désigne le groupe symétrique). Pour un module instable $M$, l'espace vectoriel $\mathrm{Hom}_{\mathcal{U}}(M, K(1)^{\otimes n})$ est muni d'une action du groupe $\mathfrak{S}_n$. Le groupe agit par permutation des facteurs sur $K(1)^{\otimes n}$. Cela induit une action à gauche de $\mathfrak{S}_n$ sur $\mathrm{Hom}_{\mathcal{U}}(M, K(1)^{\otimes n})$. Donc le dual $\mathrm{Hom}_{\mathcal{U}}(M, K(1)^{\otimes n})^{\#}$ est un $\mathbb{F}_2[\mathfrak{S}_n]$-module à droite. Donc

$$M \longmapsto \mathrm{Hom}_{\mathcal{U}}(M, K(1)^{\otimes n})^{\#}$$

détermine un foncteur $p_n : \mathcal{U} \to \mathcal{M}od_{\mathbb{F}_2[\mathfrak{S}_n]}$ (il faut prendre un dual continu). Le foncteur $p_n$ commute aux colimites et est exact à droite. Le module instable $K(1)^{\otimes n}$ étant injectif, $p_n$ est exact à gauche. Donc $p_n$ est exact. Ce foncteur est étroitement associé au $n$-ième effet croisé sur les foncteurs (voir [**7**, Section 3.1]) ; il s'y identifie pour les foncteurs de degré inférieur ou égal à $n$.

Il est trivial sur la sous-catégorie $\mathcal{N}il$ et factorise comme suit :

$$\mathcal{U} \longrightarrow \mathcal{U}/\mathcal{N}il \longrightarrow \mathcal{M}od_{\mathbb{F}_2[\mathfrak{S}_n]}.$$

Ceci définit un foncteur

$$p_n : \mathcal{U}/\mathcal{N}il \longrightarrow \mathcal{M}od_{\mathbb{F}_2[\mathfrak{S}_n]}.$$

Soit alors $\mathcal{V}_n$ la sous-catégorie pleine de $\mathcal{U}/\mathcal{N}il$ dont les objets sont ceux qui ont une image triviale par le foncteur induit par $p_k$, pour $k > n$. Par définition, le foncteur $p_n$ est trivial sur $\mathcal{V}_{n-1}$

**Théorème 13.1**. — *Le foncteur $p_n$ induit une équivalence de catégories*

$$\mathcal{V}_n/\mathcal{V}_{n-1} \longrightarrow \mathcal{M}od_{\mathbb{F}_2}[\mathfrak{S}_n].$$

Les propositions suivantes sont les principales étapes de la démonstration :

**Proposition 13.2**. — *La représentation $p_n(F(1)^{\otimes n})$ est isomorphe à $\mathbb{F}_2[\mathfrak{S}_n]$.*

**Proposition 13.3**. — *L'anneau $\mathrm{End}_{\mathcal{U}}(F(1)^{\otimes n})$ est isomorphe à $\mathbb{F}_2[\mathfrak{S}_n]$.*

**13.2. Les objets simples.** — Cette sous-section donne quelques détails sur les objets simples de $\mathcal{U}/\mathcal{N}il$ dans le cas $p = 2$. Les $\mathbb{F}_2$-représentations simples du groupe symétrique $\mathfrak{S}_n$ sont (à isomorphisme près) indexée par les partitions $\lambda = (\lambda_1, \ldots, \lambda_d)$ of $n$ telles que $\lambda_i > \lambda_{i+1}$, $1 \leqslant i \leqslant d - 1$. De telles partitions sont dites 2-régulières. La partition conjugué de $\lambda$, notée $\lambda'$, est telle que $1 \geqslant \lambda_i' - \lambda_{i+1}' \geqslant 0$, *i.e.* elle est 2-régulière pour les colonnes. Soit $R_\lambda$ la $\mathbb{F}_2$-représentation simple de $\mathfrak{S}_n$ indexée par $\lambda$. Elle est isomorphe à $\varepsilon_\lambda \mathbb{F}_2[\mathfrak{S}_n]$ pour un certain $\varepsilon_\lambda \in \mathbb{F}_2[\mathfrak{S}_n]$ qui peut être exprimé à l'aide du diagramme de Young associé à $\lambda$ [**21**]. L'objet simple $S_\lambda \in \mathcal{U}/\mathcal{N}il$ associé est de la forme $\varepsilon_\lambda F(1)^{\otimes n}$. Le foncteur simple associé $S_\lambda$ est, lui, de la forme :

$$V \longmapsto \varepsilon_\lambda V^{\otimes n}.$$

Par exemple, la représentation triviale de dimension 1 de $\mathfrak{S}_n$ (la signature si $p > 2$) correspond à la partition $(n)$ et au foncteur $n$-ième puissance extérieure.

## Références

[1] J.F. ADAMS, J.H. GUNAWARDENA & H.R. MILLER – « The Segal conjecture for elementary abelian $p$-groups », *Topology* **24** (1985), p. 435–460.

[2] J.F. ADAMS & C.W. WILKERSON – « Finite H-spaces and algebras over the Steenrod algebra », *Ann. of Math.* **111** (1980), p. 95–143.

[3] S. BULLETT & I.G. MACDONALD – « On the Adem relations », *Topology* **21** (1982), p. 329–332.

[4] H.E.A. CAMPBELL & S.P. SELICK – « Polynomial algebras over the Steenrod algebra », *Comment. Math. Helv.* **65** (1990), p. 171–180.

[5] G. CARLSSON – « G.B. Segal's Burnside ring conjecture for $(\mathbb{Z}/2)^k$ », *Topology* **22** (1983), p. 83–103.

[6] V. FRANJOU, E.M. FRIEDLANDER, A. SCORICHENKO & A. SUSLIN – « General linear and functor cohomology over finite fields », *Ann. of Math. (2)* **150** (1999), p. 663–728.

[7] V. FRANJOU & T. PIRASHVILI – « Stable $K$-theory is bifunctor homology (after A. Scorichenko) », dans ce volume.

[8] V. FRANJOU & L. SCHWARTZ – « Reduced unstable $\mathcal{A}$-modules and the modular representation theory of the symmetric groups », *Ann. scient. Éc. Norm. Sup. 4$^e$ série* **23** (1990), p. 593–624.

[9] P. GABRIEL – « Des catégories abéliennes », *Bull. Soc. math. France* **90** (1962), p. 323–348.

[10] H.-W. HENN, J. LANNES & L. SCHWARTZ – « The categories of unstable modules and unstable algebras over the Steenrod algebra modulo nilpotents objects », *Amer. J. Math.* **115** (1993), no. 5, p. 1053–1106.

[11] N. KUHN – « Generic representations of the finite general linear groups and the Steenrod algebra II », *K-Theory* **8** (1994), p. 395–426.

[12] _____, « Generic representations of the finite general linear groups and the Steenrod algebra III », *K-Theory* **9** (1995), p. 273–303.

[13] J. LANNES – « Sur les espaces fonctionnels dont la source est le classifiant d'un $p$-groupe abélien élémentaire », *Publ. Math. Inst. Hautes Études Sci.* **75** (1992), p. 135–244.

[14] J. LANNES & L. SCHWARTZ – « Sur la structure des $\mathcal{A}$-modules instables injectifs », *Topology* **28** (1989), p. 153–169.

[15] J. LANNES & S. ZARATI – « Sur les $\mathcal{U}$-injectifs », *Ann. scient. Éc. Norm. Sup. 4$^e$ série* **19** (1986), p. 1–31.

[16] _____, « Sur les foncteurs dérivés de la déstabilisation », *Math. Z.* **194** (1987), p. 25–59.

[17] H. MARGOLIS – *Spectra and the Steenrod algebra*, North Holland, 1983.

[18] W.S. MASSEY & F.P. PETERSON – *The mod. 2 cohomology structure of certain fibre spaces*, Mem. Amer. Math. Soc., vol. 74, American Mathematical Society, Providence, RI, 1967.

[19] H.R. MILLER – « The Sullivan conjecture on maps from classifying spaces », *Ann. of Math. (2)* **120** (1984), p. 39–87, Erratum : *Ibid.* **121** (1985), p. 605-609.

[20] T. PIRASHVILI – « Introduction to functor homology », dans ce volume.

[21] L. SCHWARTZ – *Unstable modules over the Steenrod algebra and Sullivan's fixed point set conjecture*, Chicago Lectures in Mathematics Series, 1994.

*Panoramas & Synthèses*
**16**, 2003, p. 101–106

# L'ALGÈBRE DE STEENROD EN TOPOLOGIE

*par*

Lionel Schwartz

*Résumé.* — On rappelle dans cette note la construction de l'algèbre de Steenrod en théorie de l'homotopie comme algèbre d'opérations naturelles stables. On donne ses principales propriétés.

*Abstract* **(The Steenrod algebra in topology).** — One recalls in this note the construction of the Steenrod algebra in homotopy theory as algebra of natural stable transformations. One gives its natural properties.

On présente ici l'algèbre de Steenrod du point de vue de la théorie de l'homotopie. On rappelle les motivations essentielles qui, du point de vue topologique, ont mené à son introduction. Ses propriétés algébriques, et ses catégories de modules et d'algèbres sont étudiées en détails dans l'article qui précède.

## 1. L'algèbre de Steenrod comme algèbre des opérations cohomologiques stables

**1.1. La cohomologie singulière.** — Soit $p$ un nombre premier. La cohomologie singulière modulo $p$ est un foncteur contravariant de la catégorie des paires $(X, A)$ $A \subset X$ d'espaces topologiques dans la catégorie **Vect**$^{\mathrm{gr}}$ des $\mathbb{F}_p$-espaces vectoriels $\mathbb{N}$-gradués :

$$(X, A) \longmapsto \mathrm{H}^n(X, A; \mathbb{Z}/p\mathbb{Z})_{n \geqslant 0}.$$

On abrègera dans la suite par $\mathrm{H}^*(X, A)$, et par $\mathrm{H}^*X$ si $A$ est l'ensemble vide. La cohomologie *réduite* $\widetilde{\mathrm{H}}^n X$ est le noyau de l'application canonique : $\mathrm{H}^n X \to \mathrm{H}^n \mathrm{pt}$.

Pourvu qu'on se restreigne à des paires d'espaces raisonnables ce foncteur vérifie les axiomes d'Eilenberg-Steenrod dont la liste suit :

---

*Classification mathématique par sujets* **(2000).** — 55S10.
*Mots clefs*. — Algèbre de Steenrod, théorie de l'homotopie.

(1) l'invariance par homotopie : deux applications $f_0, f_1 : X \to Y$ sont homotopes si il existe $F : X \times [0,1] \to Y$ telle que $F(x,0) = f_0$ et $F(x,1) = f_1$ ; on demande alors qu'elles induisent le même morphisme $f_0^* = f_1^*$ en cohomologie ;

(2) l'axiome d'excision, ou ce qui lui est équivalent l'axiome de suspension :

$$\widetilde{H}^n X \cong \widetilde{H}^{n+1} \Sigma X.$$

On rappelle que la suspension $\Sigma X$ d'un espace pointé $(X, x_0)$ est le quotient de $X \times [0,1]$ par la relation qui identifie les points $(x,0)$, $(x,1)$, $(x_0, t)$ en un seul point ; en particulier la suspension d'une sphère pointée de dimension $n$, $\Sigma S^n$, est homéomorphe à une sphère de dimension $n+1$, $S^{n+1}$ ;

(3) la longue suite exacte associée à une paire d'espaces $(X, A)$ :

$$\cdots \longrightarrow H^n(X, A) \longrightarrow H^n X \longrightarrow H^n A \longrightarrow H^{n+1}(X, A) \longrightarrow \cdots ;$$

(4) l'axiome de dimension qui donne la valeur prise par le foncteur sur les sphères : $H^i S^n \cong \mathbb{F}_p$ si $i = 0, n$, et $H^i S^n \cong \{0\}$ sinon.

S. Eilenberg et N.E. Steenrod ont montré que ces propriétés déterminent de manière unique le foncteur sur des sous-catégories d'espaces topologiques raisonnables tels que les complexes finis.

## 1.2. Le théorème de représentabilité de Brown et les espaces d'Eilenberg-MacLane.

— Fixons un entier $n$. Le foncteur $H^n$, de la catégorie des CW-complexes dans celle des groupes abéliens, vérifie les deux premiers axiomes d'Eilenberg-Steenrod. Le second axiome entraîne la propriété de Mayer-Vietoris (pour des paires assez régulières), qui s'énonce comme suit. *Soient $j_0 : A \to A \cup B$ et $j_1 : B \to A \cup B$, $f_0 : A \cap B \to A$ et $f_1 : A \cap B \to B$ les inclusions. Si $u_0 \in H^n A$ et $u_1 \in H^n B$ satisfont $f_0^*(u) = f_1^*(u)$, alors il existe $v \in H^n(A \cup B)$ tel que $j_0^*(v) = u_0$ et $j_1^*(v) = u_1$.*

De plus on a une bijection :

$$H^n(\bigvee_\lambda X_\lambda) \cong \prod_\lambda H^n(X_\lambda).$$

Ici $\bigvee_\lambda X_\lambda$ désigne le bouquet des espaces pointés $X_\lambda$, le bouquet de deux espaces pointés $(X, x_0)$ et $(Y, y_0)$ est le quotient de $X \cup Y$ par l'identification de $x_0$ à $y_0$.

Le théorème de représentabilité de Brown (voir [**2**, §VII.7]) implique alors qu'il existe un espace $K_n$ tel que le foncteur $H^n(-)$ et le foncteur $[-, K_n]$ des classes d'homotopie d'applications vers $K_n$ soient naturellement équivalents. L'équivalence est décrite comme suit : le groupe $H^n K_n$ est isomorphe à $\mathbb{F}_p$, on en choisit un générateur $\iota_n$. La transformation naturelle associe à la classe d'homotopie d'une l'application $f$ l'élément $f^*(\iota_n)$.

Par construction, et du fait de la valeur de la cohomologie des sphères, l'espace $K_n$ a tous ses groupes d'homotopie triviaux, sauf le $n$-ième qui est isomorphe à $\mathbb{F}_p$ : c'est ce qu'on appelle un espace d'Eilenberg-MacLane $K(\mathbb{F}_p, n)$.

## 1.3. Les opérations cohomologiques, les relations d'Adem. — Si on se donne une application

$$\theta_n : \ K_n \longrightarrow K_{n+r}$$

on en déduit par composition une application naturelle

$$\theta_n^* : \ [X, K_n] \cong \mathrm{H}^n X \longrightarrow [X, K_{n+r}] \cong \mathrm{H}^{n+r} X.,$$

qui ne dépend que la classe d'homotopie de $\theta_n$. Inversement une transformation naturelle est entièrement déterminée par une telle application. En effet, si on s'est donné une telle transformation naturelle, l'application associée est l'image de l'identité de $K(\mathbb{F}_p, n)$.

On appelle opération naturelle stable de degré $r$ de la cohomologie singulière modulo $p$ la donnée pour tout entier $n$ d'une transformation naturelle de foncteurs :

$$t_n : \ \mathrm{H}^n(-) \longrightarrow \mathrm{H}^{n+r}(-),$$

satisfaisant à la condition de stabilité suivante :

$$t_n(\Sigma x) = \Sigma(t_{n+1}(x)),$$

pour tout $x \in \widetilde{\mathrm{H}}^n X$. Dans la formule ci-dessus, conformément à l'usage, si $x \in \widetilde{\mathrm{H}}^n X$ on note $\Sigma x \in \widetilde{\mathrm{H}}^{n+1}(\Sigma X)$ la classe correspondante par l'isomorphisme de l'axiome de suspension. L'ensemble de ces opérations forme une $\mathbb{F}_p$-algèbre graduée, le produit étant donné par la composition.

Ainsi qu'on l'a dit, la donnée d'une telle famille est équivalente à celle d'une famille d'applications $\theta_n : K_{n+r} \to K_n$ satisfaisant à une propriété *ad hoc*.

L'ensemble de ces opérations avec la somme et la composition comme produit constitue une $\mathbb{F}_p$-algèbre graduée, que l'on notera $\mathcal{A}_p$ et qui s'appelle *l'algèbre de Steenrod*.

Avec cette définition on a évidemment :

**Théorème 1.1**. — *Pour tout espace $X$, $\mathrm{H}^* X$ est naturellement un module sur l'algèbre de Steenrod $\mathcal{A}_p$.*

Décrivons cette algèbre. Soit la $\mathbb{F}_p$-algèbre graduée associative unitaire engendrée par des éléments $\widetilde{\mathrm{Sq}}^i$ de degré $i$ (resp. $\widetilde{\mathrm{P}}^i$ de degré $2i(p-1)$, $i > 0$, et $\widetilde{\beta}$ de degré 1 tel que $\widetilde{\beta}^2 = 0$). On considère son quotient par les relations suivantes, dites relations d'Adem :

– pour $p = 2$ :

$$\widetilde{\mathrm{Sq}}^a \widetilde{\mathrm{Sq}}^b = \sum_0^{[a/2]} \binom{b-j-1}{a-2j} \widetilde{\mathrm{Sq}}^{a+b-j} \widetilde{\mathrm{Sq}}^j$$

pour tous $a, b > 0$ (la restriction $a < 2b$ n'est pas nécessaire) ;

– pour $p > 2$ :

$$\widetilde{\mathrm{P}}^a \widetilde{\mathrm{P}}^b = \sum_0^{[a/p]} (-1)^{a+j} \binom{(p-1)(b-j)-1}{a-pj} \widetilde{\mathrm{P}}^{a+b-j}\widetilde{\mathrm{P}}^j$$

$$\widetilde{\mathrm{P}}^a \widetilde{\beta}\widetilde{\mathrm{P}}^b = \sum_0^{[a/p]} (-1)^{a+i} \binom{(p-1)(b-i)}{a-pt} \widetilde{\beta}\widetilde{\mathrm{P}}^{a+b-i}\widetilde{\mathrm{P}}^i$$

$$+ \sum_0^{[(a-1)/p]} (-1)^{a+i-1} \binom{(p-1)(b-i)-1}{a-pi-1} \widetilde{\mathrm{P}}^{a+b-i}\widetilde{\beta}\widetilde{\mathrm{P}}^t$$

pour tous $a, b > 0$ (la condition $a < pb$ n'est pas nécessaire).

N.E. Steenrod a construit des opérations cohomologiques stables $\mathrm{Sq}^i$ et $\mathrm{P}^i$ qui agissent sur la cohomologie singulière, puis J. Adem a montré que les relations ci-dessus ont lieu dans la cohomologie mod $p$ de n'importe quel espace, et qu'elles engendrent l'idéal des éléments agissant trivialement [3]. Le quotient ci-dessus est donc une sous-algèbre de l'algèbre des opérations stables. Le calcul par H. Cartan et J.-P. Serre [1] de la cohomologie des espaces d'Eilenberg-Mac Lane montre que ces deux algèbres sont isomorphes.

## 2. La condition d'instabilité et les algèbres instables

**2.1. Modules instables.** — La cohomologie modulo $p$ d'un espace $X$ est un $\mathcal{A}_p$-module d'un type particulier : il est instable.

***Définition 2.1***. — Un $\mathcal{A}_p$-module $M$ est dit instable si il satisfait la condition suivante.
  – pour $p = 2$ : si $x \in \mathrm{H}^*X$, et $i > |x|$, alors $\mathrm{Sq}^i x = 0$ ;
  – pour $p > 2$ : si $x \in \mathrm{H}^*X$, et $e + 2i > |x|$, $e = 0, 1$, alors $\beta^e P^i x = 0$.

Ici $|x|$ désigne le degré de $x$. En particulier, un $\mathcal{A}_p$-module instable $M$ est trivial en degrés négatifs. On écrit modulo instable au lieu de $\mathcal{A}_p$-module instable la plupart du temps.

La sous-catégorie pleine de la catégorie des $\mathcal{A}_p$-modules dont les objets sont les modules instables est notée $\mathcal{U}$ ; elle est abélienne.

**2.2. Algèbres instables.** — La cohomologie modulo $p$ d'un espace $X$, munie du cup-produit, est une $\mathbb{F}_p$-algèbre $\mathbb{N}$-graduée, commutative, unitaire, et ce naturellement. Cette structure est reliée à la structure de $\mathcal{A}_p$-module par deux propriétés :

  $(\mathcal{K}_1)$ la « formule de Cartan » :
  – si $p = 2$

$$\mathrm{Sq}^i(xy) = \sum_{k+\ell=i} \mathrm{Sq}^k x\, \mathrm{Sq}^\ell y;$$

– si $p > 2$ :

$$P^i(xy) = \sum_{k+\ell=i} P^k x \, P^\ell y,$$

$$\beta(xy) = (\beta x)\, y + (-1)^{|x|} x \, \beta y$$

pour tous $x, y \in \mathrm{H}^* X$.

$(\mathcal{K}_2)$

– si $p = 2$ : $\mathrm{Sq}^{|x|} x = x^2$, pour tout $x \in \mathrm{H}^* X$ ;

– si $p > 2$ : $P^{|x|/2} x = x^p$ pour tout $x$ de degré pair dans $\mathrm{H}^* X$.

Ceci conduit à la :

**Définition 2.2**. — Une $\mathcal{A}_p$-algèbre instable $K$ est un module instable muni d'une structure de $\mathbb{F}_p$-algèbre commutative, unitaire, et dont le produit vérifie les propriétés $\mathcal{K}_1$ et $\mathcal{K}_2$.

On note $\mathcal{K}$ la catégorie des $\mathcal{A}_p$-algèbres instables, dont les morphismes sont les applications d'algèbres, $\mathcal{A}_p$-linéaires, de degré zero. On dit *algèbre instable* au lieu de $\mathcal{A}_p$-algèbre instable.

**Exemple 2.3**. — La cohomologie modulo 2 du groupe $\mathbb{Z}/2$, $\mathrm{H}^*(\mathbb{Z}/2)$, est une algèbre polynomiale $\mathbb{F}_2[u]$ en un générateur $u$ de degré 1. L'action de $\mathcal{A}_2$ est complètement déterminée par $\mathcal{K}_1$ et $\mathcal{K}_2$. On trouve :

$$\mathrm{Sq}^i u^n = \binom{n}{i} u^{n+i}.$$

Notons que cette cohomologie est aussi la cohomologie de l'espace classifiant $B\mathbb{Z}/2$, et que ce dernier n'est autre que l'espace projectif infini $\mathbb{R}P^\infty$.

Pour $p > 2$, $\mathrm{H}^*(\mathbb{Z}/p)$ est le produit tensoriel $E(t) \otimes \mathbb{F}_p[x]$ d'une algèbre extérieure en un générateur $t$ de degré 1 et d'une algèbre polynomiale en un générateur $x$ de degré 2. L'action de $\mathcal{A}_p$ est determinée par $\mathcal{K}_1$, $\mathcal{K}_2$ et le fait que $\beta$ est l'homomorphisme de Bockstein. On obtient :

$$\beta t = x, \quad P^i x^n = \binom{n}{i} x^{n+i(p-1)}.$$

**Exemple 2.4**. — Pour $p = 2$, le sous-$\mathcal{A}_2$-module de $\mathrm{H}^* B\mathbb{Z}/2$ engendré par $u$ est noté $F(1)$. L'ensemble $\{u, u^2, u^4, \ldots\}$ est une base de $F(1)$ comme $\mathbb{F}_2$-espace vectoriel gradué.

Pour $p > 2$, le sous-$\mathcal{A}_p$-module de $\mathrm{H}^* B\mathbb{Z}/p$ engendré par $t$ est noté $F(1)$. L'ensemble $\{t, x, x^p, x^{p^2}, \ldots\}$ est une de base $F(1)$ comme $\mathbb{F}_p$-espace vectoriel gradué.

## Références

[1] J.-P. SERRE – « Cohomologie modulo 2 des complexes d'Eilenberg-Mac-Lane », *Comment. Math. Helv.* **27** (1953), p. 198–232.

[2]  E. Spanier – *Algebraic Topology*, Springer Verlag (originally published by Mc Graw-Hill), 1966.

[3]  N.E. Steenrod & D.B.A. Epstein – *Cohomology operations*, Annals of Math. Studies, vol. 50, Princeton University Press, Princeton, NJ, 1962.

*Panoramas & Synthèses*
**16**, 2003, p. 107–126

# STABLE K-THEORY IS BIFUNCTOR HOMOLOGY
## (AFTER A. SCORICHENKO)

*by*

Vincent Franjou & Teimuraz Pirashvili

**Abstract.** — For many rings $R$, the homology with coefficients of the infinite general linear group $\mathrm{GL}(R)$ is the tensor product of its homology with trivial coefficients with another term, which has been identified as the stable $K$-theory of the ring. Scorichenko's theorem states that stable $K$-theory is functor homology.

***Résumé* (La K-théorie stable est l'homologie des foncteurs (d'après A. Scorichenko))**
    Pour beaucoup d'anneaux $R$, l'homologie du groupe linéaire infini avec coefficients s'obtient en effectuant le produit tensoriel de son homologie avec coefficients triviaux par un autre terme, qui n'est autre que la $K$-théorie stable de l'anneau. Le théorème de Scorichenko exprime la $K$-théorie stable comme homologie des foncteurs.

## 0. Introduction

The purpose of this chapter is to present A. Scorichenko's work for his dissertation at Northwestern.

**Theorem 0.1 ([20]).** — *For a ring $R$, let $\mathbb{P}(R)$ be the category of finitely generated projective left $R$-modules, and let $D : \mathbb{P}(R)^{\mathrm{op}} \times \mathbb{P}(R) \to \mathrm{Ab}$ be a bifunctor. If $D$ has finite degree with respect to both variables, then there is an isomorphism between Waldhausen's stable K-theory and the homology of $\mathbb{P}$:*

$$K_*^{\mathrm{st}}(R, D) \longrightarrow \mathrm{H}_*(\mathbb{P}(R), D).$$

This proves a conjecture stated in [**4**]. The conjecture first appeared in [**15**] for biadditive bifunctors, a case proved in [**6**] (see also [**19**] for the outline of another approach). In the case of finite fields, the conjecture was proved for general bifunctor coefficients in [**3**] and in [**8**, Appendix]. This special case can be reformulated in terms of functor cohomology, whose computation is a main topic in this book. Indeed, the

---

**2000** *Mathematics Subject Classification*. — 19D55.
**Key words and phrases**. — K-theory, general linear group, polynomial functor, bifunctor, MacLane homology, homology of a small category, spectral sequences.

conjecture has been a motivation for developing computation tools in categories of functors. The homology $H_*(\mathbb{P}(R), D)$ can be expressed purely in terms of homological algebra in categories of functors, which is well understood in many cases (see [**9, 10, 8**] or the article *Introduction to functor homology* in this volume). For example, when $D(X, Y) = \mathrm{Hom}_R(X, P \otimes_R Y)$, $H_*(\mathbb{P}(R), D)$ is isomorphic to the topological Hochschild homology [**16**] and to the MacLane homology [**11**] of $R$ with coefficients in the bimodule $P$. When $D(X, Y) = \mathrm{Hom}_R(A(X), B(Y))$ for polynomial functors $A$ and $B$ in $\mathcal{F}(R)$, and if $R$ is a field, then $H_*(\mathbb{P}(R), D)$ is dual to $\mathrm{Ext}^*_{\mathcal{F}(R)}(A, B)$ as studied in this book.

Stable K-theory is precisely related to homology of invertible matrices: Waldhausen explained [**21**, Section 6] that stable K-theory gives access to homology of the general linear group, with twisted coefficients, through the spectral sequence discussed in sections 1 and 5. One point of the theorem is that although stable K-theory is defined in terms of invertible matrices, it is naturally isomorphic to a more manageable theory, expressed in terms of all matrices. The isomorphism of Theorem 0.1 is induced by the inclusion of invertible matrices in all matrices. There are variations on this, as will be seen with Scorichenko's use of the category of epimorphisms.

## 1. Homology of general linear groups and stable K-theory

Let $R$ be a ring and $\mathrm{GL}_n(R)$ be the group of invertible matrices over $R$. For a bimodule $P$ over $R$, the $R$-bimodule of $n \times n$-matrices $\mathrm{gl}_n(P)$ is a $\mathrm{GL}_n(R)$-module for the conjugation action: $X * M := X^{-1}MX$. We embed $\mathrm{GL}_n(R)$ as a subgroup in $\mathrm{GL}_{n+1}(R)$ by: $X \mapsto \left( \begin{smallmatrix} X & 0 \\ 0 & 1 \end{smallmatrix} \right)$, and define the direct limit $\mathrm{GL}(R) = \bigcup_n \mathrm{GL}_n(R)$. We embed $\mathrm{gl}_n(P)$ in $\mathrm{gl}_{n+1}(P)$ by: $M \mapsto \left( \begin{smallmatrix} M & 0 \\ 0 & 0 \end{smallmatrix} \right)$, and define the direct limit $\mathrm{gl}(P) = \bigcup_n \mathrm{gl}_n(P)$. This yields the conjugation action of $\mathrm{GL}(R)$ on $\mathrm{gl}(P)$.

The homology groups with twisted coefficients $H_*(\mathrm{GL}(R); \mathrm{gl}(P))$ appear as the $E^2_{n1}$-terms of the following change of rings spectral sequence. Let:

$$0 \longrightarrow P \longrightarrow S \longrightarrow R \longrightarrow 0$$

be a singular extension of rings. Thus $S$ is a ring and $P$ is a two-sided ideal of $S$ such that $P^2 = 0$ and $R = S/P$. There is a short exact sequence of groups

$$0 \longrightarrow \mathrm{gl}(P) \longrightarrow \mathrm{GL}(S) \longrightarrow \mathrm{GL}(R) \longrightarrow 1,$$

where the inclusion is by the exponential map $x \mapsto 1 + x$. It yields a Hochschild-Serre spectral sequence

$$E^2_{pq} = H_p(\mathrm{GL}(R), H_q(\mathrm{gl}(P))) \Longrightarrow H_{p+q}(\mathrm{GL}(S)).$$

Since $\mathrm{gl}(P)$ is an abelian group, its homology $H_*(\mathrm{gl}(P))$ is known [**18**, Section 8]. Here is a way to put these groups in a more general framework.

Let $\mathbb{P}(R)$, or simply $\mathbb{P}$, be the category of finitely generated projective left $R$-modules. The category $\mathbb{P}$ is equivalent to a small category and therefore we can do homological algebra in $\mathrm{Func}(\mathbb{P}, \mathrm{Ab})$. For a bifunctor $D : \mathbb{P}^{\mathrm{op}} \times \mathbb{P} \to \mathrm{Ab}$ the abelian group $D(R^n, R^n)$ has a natural $\mathrm{GL}_n(R)$-module structure, with action on both variable. Define $p_n : R^{n+1} \to R^n$ and $i_n : R^n \to R^{n+1}$ by: $p_n(x_1, \ldots, x_{n+1}) = (x_1, \ldots, x_n)$ and $i_n(x_1 \ldots, x_n) = (x_1, \ldots, x_n, 0)$. They yield an homomorphism

$$D(p_n, i_n) : D(R^n, R^n) \longrightarrow D(R^{n+1}, R^{n+1})$$

which is compatible with the inclusions $\mathrm{GL}_n(R) \subset \mathrm{GL}_{n+1}(R)$. At the limit, one gets a $\mathrm{GL}(R)$-module $D_\infty := \mathrm{colim}_n D(R^n, R^n)$. For example, when $D(X, Y) = \mathrm{Hom}_R(X, P \otimes_R Y)$ for a given bimodule $P$, then $D_\infty = \mathrm{gl}(P)$. Considering the bifunctor defined by $D(X, Y) = \mathrm{H}_q(\mathrm{Hom}_R(X, P \otimes_R Y))$ recovers $D_\infty = \mathrm{H}_q(\mathrm{gl}(P))$.

We are left with the general problem of understanding the groups $\mathrm{H}_*(\mathrm{GL}(R), D_\infty)$. This is achieved by comparing it with an appropriate notion of homology of a small category for the category $\mathbb{P}$ (see Section 2.5). The group $\mathrm{GL}_n(R)$ appears as the subcategory of $\mathbb{P}$ consisting of the automorphisms of $R^n$, and this inclusion induces an homomorphism

$$\psi_* : \mathrm{H}_*(\mathrm{GL}(R), D_\infty) \longrightarrow \mathrm{H}_*(\mathbb{P}(R), D).$$

Unfortunately the homomorphism $\psi_*$ is very far from being an isomorphism in general. Indeed, if $D$ is a constant bifunctor, then $\mathrm{H}_*(\mathbb{P}, D)$ vanishes in positive dimensions, because $\mathbb{P}$ has a zero object, while the homology of the general linear group is highly nontrivial in general. There is a trick due to Waldhausen [21, p. 387–388], which simplifies the situation. Define the stable $K$-theory $K_*^{\mathrm{st}}(R, D)$ of $R$ with coefficients in $D$ as the homology of the homotopy fiber of $B\,\mathrm{GL}(R) \to B\,\mathrm{GL}(R)^+$, with twisted coefficients in $D_\infty$. In the resulting Serre spectral sequence

$$(1) \qquad E_{pq}^2 = \mathrm{H}_*(\mathrm{GL}(R), K_*^{\mathrm{st}}(R, D)) \Longrightarrow \mathrm{H}_*(\mathrm{GL}(R), D_\infty)$$

the action of $\mathrm{GL}(R)$ on $K_*^{\mathrm{st}}(R, D)$ is *trivial* (see [12]). The spectral sequence (1) degenerates at $E^2$ in many cases (see [4], or Section 5 in this paper). Moreover there is a natural transformation

$$\nu_* : K_*^{\mathrm{st}}(R, D) \longrightarrow \mathrm{H}_*(\mathbb{P}(R), D)$$

because $\mathrm{H}_*(\mathbb{P}, -)$ is a universal sequence of functors defined on $\mathrm{Func}(\mathbb{P}^{\mathrm{op}} \times \mathbb{P}, \mathrm{Ab})$ (see Lemma 2.1).

Scorichenko's theorem 0.1 states that $\nu_*$ is an isomorphism, if $D$ has finite degree with respect to both variables. For the definition of functors of finite degree we refer the reader to Section 3. Symmetric, exterior or divided powers all have finite degree, as does indeed the bifunctor defined by $D(X, Y) = \mathrm{H}_q(\mathrm{Hom}_R(X, P \otimes_R Y))$, which is relevant to the above change of rings spectral sequence.

## 2. Preliminaries from homological algebra

**2.1. Universal sequences of functors.** — We assume the reader to be familiar with the basics of homological algebra and category theory, as in [**5**]. We recall the following axiomatic characterization of derived functors, to be used several times in this paper. Let **A** and **B** be abelian categories. A *connected sequence of functors* is a sequence of additive functors $(T_n : \mathbf{A} \to \mathbf{B})_{n \geqslant 0}$ together with homomorphisms

$$\partial_n : T_{n+1}(C) \longrightarrow T_n(A)$$

for each exact sequence in **A**

$$0 \longrightarrow A \xrightarrow{i} B \xrightarrow{s} C \longrightarrow 0$$

which are natural in respect of maps of short exact sequences. A connected sequence is *exact* if for each exact sequence $0 \to A \xrightarrow{i} B \xrightarrow{s} C \to 0$ in **A**, the long sequence in **B**

$$\cdots \longrightarrow T_{n+1}(C) \xrightarrow{\partial} T_n(A) \xrightarrow{i_*} T_n(B) \xrightarrow{s_*} T_n(C) \longrightarrow \cdots \longrightarrow T_0(C) \longrightarrow 0$$

is exact. Assume **A** has enough projective objects. A *universal sequence of functors* is an exact connected sequence of functors such that $T_n(P) = 0$ for all positive $n$ and all projective $P$. The following is a particular case of [**5**, Proposition III.5.2].

**Proposition 2.1.** — *Let $T : \mathbf{A} \to \mathbf{B}$ be an additive covariant functor. Its left derived functors $(L_n T : \mathbf{A} \to \mathbf{B})_{n \geqslant 0}$ form a universal sequence of functors. Conversely, if $(T_n : \mathbf{A} \to \mathbf{B})_{n \geqslant 0}$ is an exact connected sequence of functors, then there is a unique morphism of connected sequence of functors $(\xi_n : T_n \to L_n(T_0))_{n \geqslant 0}$ such that $\xi_0 : T_0 \to L_0 T_0$ is the canonical isomorphism. Furthermore $\xi_n$ is an isomorphism for all $n \geqslant 0$ provided $(T_n : \mathbf{A} \to \mathbf{B})_{n \geqslant 0}$ is a universal sequence of functors.*

**2.2. A lemma on collapsing spectral sequences.** — We now extend these notions to spectral sequences of functors. A $\partial$-*spectral sequence* is for each $A \in \mathbf{A}$ a upper-half-plane spectral sequence $(E^r_{pq}(A), d^r)_{r \geqslant 2}$ in **B**, which is natural in $A \in \mathbf{A}$, together with homomorphisms

$$\partial_r : E^r_{pq}(C) \longrightarrow E^r_{p,q-1}(A)$$

for each short exact sequence $0 \to A \xrightarrow{i} B \xrightarrow{s} C \to 0$ in **A**, which are natural in respect of maps of short exact sequences, and such that:

(i) for each $r \geqslant 2$, $\partial_{r+1}$ is the map induced in homology by $\partial_r$
(ii) the diagrams

$$\begin{array}{ccc} E^r_{pq}(C) & \xrightarrow{\ d^r\ } & E^r_{p-r,q+r-1}(C) \\ {\scriptstyle \partial}\downarrow & & \downarrow{\scriptstyle \partial} \\ E^r_{p,q-1}(A) & \xrightarrow{\ d^r\ } & E^r_{p-r,q+r-2}(A) \end{array}$$

commute for all integers $p$, $q$, and $r \geqslant 2$.

**Lemma 2.2**. — *Let* $\mathbf{A}$ *be an abelian category and let* $(E^r_{pq})_{r \geqslant 2}$ *be a* $\partial$-*spectral sequence. Assume that the following condition holds: For any* $C$ *in* $\mathbf{A}$, *there is a short exact sequence* $0 \to A \to B \to C \to 0$ *in* $\mathbf{A}$ *such that the maps* $\partial^2 : E^2_{pq}(C) \to E^2_{p,q-1}(A)$ *are monomorphisms. Then the spectral sequence* $(E^r_{pq}(C), d^r)_{r \geqslant 2}$ *stops at* $E^2$ *for any* $C$ *in* $\mathbf{A}$.

*Proof.* — We need to show that $d^r = 0$ for each $r$. Let $C$ be in $\mathbf{A}$, and let $0 \to A \to B \to C \to 0$ be a short exact sequence as in the statement. Starting at the $E^2$-level, let us consider the commutative diagram:

$$
\begin{array}{ccc}
E^2_{pq}(C) & \xrightarrow{d^2} & E^2_{p-2,q+1}(C) \\
\Big\downarrow{\partial} & & \Big\downarrow{\partial} \\
E^2_{p,q-1}(A) & \xrightarrow{d^2} & E^2_{p-2,q}(A)
\end{array}
$$

By hypothesis, the right vertical map is mono. When $q = 0$, the left bottom term is 0: hence $d^2_{p0} = 0$. We then proceed by induction on $q$, applying the induction hypothesis to $A$ to show that the bottom map is 0.

At the next stage, we have $E^3 = E^2$ and $\partial_3 = \partial_2$, by the first condition of a $\partial$-spectral sequence. Hence the conditions on the $E^2$-term carry over to the $E^3$-term, and we repeat the argument *ad lib*. $\qquad\square$

**2.3. Categories of functors.** — For a small category $\mathcal{C}$ and a category $\mathbf{A}$ we let $\mathrm{Func}(\mathcal{C}, \mathbf{A})$ be the category of all functors from $\mathcal{C}$ to $\mathbf{A}$ and natural transformations between them. The category $\mathrm{Func}(\mathcal{C}, \mathbf{A})$ carries lots of the properties of $\mathbf{A}$. It has limits (resp. colimits) provided $\mathbf{A}$ has limits (resp. colimits). The limits and colimits in $\mathrm{Func}(\mathcal{C}, \mathbf{A})$ are computed pointwise. In particular, if $\mathbf{A}$ is an abelian category, then $\mathrm{Func}(\mathcal{C}, \mathbf{A})$ is also an abelian category: A sequence

$$0 \longrightarrow F \longrightarrow G \longrightarrow H \longrightarrow 0$$

is an exact sequence in $\mathrm{Func}(\mathcal{C}, \mathbf{A})$ if

$$0 \longrightarrow F(X) \longrightarrow G(X) \longrightarrow H(X) \longrightarrow 0$$

is exact for all $X \in \mathcal{C}$.

We are especially interested in the case when $\mathbf{A}$ is the category $R$-Mod of left modules over a ring $R$. We restrict to this case for the rest of the section. To describe projective generators in the category $\mathrm{Func}(\mathcal{C}, \mathbf{A})$, we recall the Yoneda lemma.

**Lemma 2.3 ([13])**. — *Let* $X$ *be an object in* $\mathcal{C}$. *For any functor*

$$T : \mathcal{C} \longrightarrow \mathrm{Sets}$$

*to the category of sets, there is a natural (in $X$) bijection*

$$\mathrm{Hom}_{\mathrm{Func}(\mathcal{C}, \mathrm{Sets})}(\mathrm{Hom}_{\mathcal{C}}(X, -), T) \ \cong \ T(X),$$

*which assigns $\xi_X(1_X) \in T(X)$ to a natural transformation $\xi : \mathrm{Hom}_{\mathcal{C}}(X, -) \to T$. Its inverse associates to each a in $T(X)$ the natural transformation $\mathrm{Hom}_{\mathcal{C}}(X, Y) \to T(Y)$ given by evaluation $f \mapsto T(f)(a)$.*

For any $X \in \mathcal{C}$, let us define $P_X \in \mathrm{Func}(\mathcal{C}, \mathbf{A})$ by

$$P_X(Y) := R[\mathrm{Hom}(X, Y)] = \bigoplus_{f:X \to Y} R.$$

Here and below $R[S]$ denotes the free left $R$-module generated by a set $S$ (it is a covariant functor of $S$). Sometimes, to emphasize the category $\mathcal{C}$ we write $P_X^{\mathcal{C}}$ instead of $P_X$.

**Corollary 2.4.** — *Let $\mathbf{A}$ be the category of left $R$-modules.*

(i) *For any $X \in \mathcal{C}$ and any functor $F : \mathcal{C} \to \mathbf{A}$ there is a natural isomorphism*

$$\mathrm{Hom}_{\mathrm{Func}(\mathcal{C}, \mathbf{A})}(P_X, F) \cong F(X).$$

(ii) *For any $X \in \mathcal{C}$, the functor $P_X$ is a projective object in $\mathrm{Func}(\mathcal{C}, \mathbf{A})$.*

(iii) *Any projective object in $\mathrm{Func}(\mathcal{C}, \mathbf{A})$ is a direct summand in a coproduct of objects $P_X$.*

(iv) *For any object $F \in \mathrm{Func}(\mathcal{C}, \mathbf{A})$ there is an epimorphism $P \to F$ with projective $P$.*

*Proof.* — The first statement is an immediate consequence of the Yoneda lemma. The functor $\mathrm{Hom}_{\mathrm{Func}(\mathcal{C}, \mathbf{A})}(P_X, -)$ is an exact functor, thanks to (i) and we obtain (ii). Take any functor $F : \mathcal{C} \to \mathbf{A}$ and an element $x \in F(X)$. Thanks to (i) we have a morphism $\xi_x : P_X \to F$ such that $(\xi_x)_X(1_X) = x$. The collection of all $(\xi_x)$, $x \in F(X)$, where $X$ runs through the isomorphism classes of objects of the category $\mathcal{C}$, yields a homomorphism

$$\xi = (\xi_x) : \bigoplus_X \bigoplus_{x \in F(X)} P_X \longrightarrow F$$

which is clearly an epimorphism. This implies (iii) and (iv). $\square$

**2.4. Tor in functor categories.** — We now discuss Tor groups in categories of functors. Assume $M : \mathcal{C} \to R\text{-Mod}$ and $N : \mathcal{C}^{\mathrm{op}} \to \text{Mod-}R$ are functors to the category of left and right $R$-modules respectively. We let $N \otimes_{\mathcal{C}} M$ be the abelian group generated by all symbols $x \otimes y$, where $x \in N(A)$, $y \in M(A)$ and $A \in \mathcal{C}$, subject to the following relations:

$$(x_1 + x_2) \otimes y = x_1 \otimes y + x_2 \otimes y, \qquad x \otimes (y_1 + y_2) = x \otimes y_1 + x \otimes y_2,$$
$$(xr) \otimes y = x \otimes (ry), \qquad \alpha^*(x') \otimes y = x' \otimes \alpha_*(y),$$

for $\alpha : A \to B$ a morphism in $\mathcal{C}$, $x_1$, $x_2$, $x$ in $N(A)$, $y_1$, $y_2$, $y$ in $M(B)$, $x'$ in $N(B)$ and $r$ in $R$. In other words: $N \otimes_{\mathcal{C}} M$ is the quotient of $\bigoplus_{A \in \mathcal{C}} N(A) \otimes_R M(A)$ by the relations $\alpha^*(x') \otimes y = x' \otimes \alpha_*(y)$. The bifunctor $- \otimes_{\mathcal{C}} -$ is right exact with respect to each variable and preserves direct sums.

**Example 2.5**. — If $R$ denotes the constant functor with value $R$, then $R \otimes_{\mathcal{C}} M$ is isomorphic to the colimit of $M : \mathcal{C} \to R$-Mod.

**Lemma 2.6**. — *For any functors $M : \mathcal{C} \to R$-Mod, $N : \mathcal{C} \to$ Mod-$R$ and any $A$ in $\mathcal{C}$, there exist natural isomorphisms*

$$N \otimes_{\mathcal{C}} P_A^{\mathcal{C}} \cong N(A)$$

$$P_A^{\mathcal{C}^{\mathrm{op}}} \otimes_{\mathcal{C}} M \cong M(A).$$

*Here as usual $P_A^{\mathcal{C}} = R[\mathrm{Hom}_{\mathcal{C}}(A, -)]$ and $P_A^{\mathcal{C}^{\mathrm{op}}} = R^{\mathrm{op}}[\mathrm{Hom}_{\mathcal{C}}(-, A)]$.*

*Proof*. — We define mutually inverse homomorphisms $f : N \otimes_{\mathcal{C}} P_A^{\mathcal{C}} \to N(A)$ and $g : N(A) \to N \otimes_{\mathcal{C}} P_A^{\mathcal{C}}$ by $f(x \otimes \alpha) = N(\alpha)(x)$ and $g(a) = a \otimes 1_A$ for $a$ in $N(A)$, $x$ in $N(X)$ and $\alpha : A \to X$ a morphism in $\mathcal{C}$. Similarly for the second isomorphism. $\square$

As usual, the left derived functors of $- \otimes_{\mathcal{C}} -$ are denoted by $\mathrm{Tor}_*^{\mathcal{C}}(-, -)$.

The following lemma is similar to a change of rings in Tor-groups. Note that any functor $f : \mathcal{C} \to \mathcal{D}$ between small categories yields a functor $f^* : \mathrm{Func}(\mathcal{D}, \mathcal{E}) \to \mathrm{Func}(\mathcal{C}, \mathcal{E})$ defined by pre-composition: $f^* R = R \circ f$.

**Lemma 2.7**. — *Let $\mathcal{C}$ and $\mathcal{D}$ be small categories. Let $l : \mathcal{C} \to \mathcal{D}$ and $r : \mathcal{D} \to \mathcal{C}$ form a pair of adjoint functors. For any $F : \mathcal{C} \to R$-Mod and $G : \mathcal{D}^{\mathrm{op}} \to$ Mod-$R$, there is an isomorphism*

$$\mathrm{Tor}_*^{\mathcal{D}}(G, r^* F) \cong \mathrm{Tor}_*^{\mathcal{C}}(l^* G, F).$$

*Proof*. — For $A \in \mathcal{C}$ and $B \in \mathcal{D}$ there are isomorphisms:

$$r^* P_A^{\mathcal{C}} \cong P_{lA}^{\mathcal{D}}, \quad \text{and} \quad l^* P_B^{\mathcal{D}^{\mathrm{op}}} \cong P_{rB}^{\mathcal{C}^{\mathrm{op}}}.$$

Therefore $l^*$ and $r^*$ respect projective objects. Furthermore one has

$$l^* P_B^{\mathcal{D}^{\mathrm{op}}} \otimes_{\mathcal{C}} P_A^{\mathcal{C}} \cong P_{rB}^{\mathcal{C}^{\mathrm{op}}} \otimes_{\mathcal{C}} P_A^{\mathcal{C}} \cong R[\mathrm{Hom}_{\mathcal{C}}(A, rB)]$$

and

$$P_B^{\mathcal{D}^{\mathrm{op}}} \otimes_{\mathcal{C}} r^* P_A^{\mathcal{C}} \cong P_B^{\mathcal{D}^{\mathrm{op}}} \otimes_{\mathcal{D}} P_{lA}^{\mathcal{D}} \cong R[\mathrm{Hom}_{\mathcal{D}}(lA, B)]$$

Thus $l^* G \otimes_{\mathcal{C}} F \cong G \otimes_{\mathcal{D}} r^* F$ as abelian groups, provided both $G$ and $F$ are projective objects. Since $\otimes$ is right exact it follows that the isomorphism exists for any $F$ and $G$. This proves the lemma in dimension 0. Since $r^*$ and $l^*$ are exact functors and send projective objects to projectives, the result is also true in all dimensions. $\square$

## 2.5. Homology of small categories.

— Let $\mathcal{C}$ be a small category and let

$$D : \mathcal{C}^{\mathrm{op}} \times \mathcal{C} \longrightarrow \mathrm{Ab}$$

be a bifunctor, which is contravariant with respect to the first argument and covariant with respect to the second argument. Thus, for $x$ in $D(X, Y)$, $f : Y \to Z$ and $g : W \to X$, one has: $f_* x \in D(X, Z)$ and $g^* x \in D(W, Y)$.

Consider diagrams in $\mathcal{C}$:

$$X_0 \xleftarrow{\ f_1\ } X_1 \xleftarrow{\ f_2\ } \cdots \xleftarrow{\ f_n\ } X_n.$$

As usual, let $\mathrm{N}_n(\mathcal{C})$ be the set of all such diagrams, which we denote, for short, by $f$ or, for $n > 0$, by $(f_1, \dots, f_n)$. Define:

$$\mathrm{F}_n(\mathcal{C}, D) := \bigoplus_{f \in \mathrm{N}_n(\mathcal{C})} D(X_0, X_n).$$

For $n = 0$,

$$\mathrm{F}_0(\mathcal{C}, D) = \bigoplus_{X_0 \in \mathcal{C}} D(X_0, X_0).$$

A typical generator of $\mathrm{F}_n(\mathcal{C}, D)$ is denoted by $(a; f_1, \dots, f_n)$, $a \in D(X_0, X_n)$. The boundary map

$$d : \mathrm{F}_n(\mathcal{C}, D) \longrightarrow \mathrm{F}_{n-1}(\mathcal{C}, D), \quad n > 0,$$

is defined by

$$d(a; f_1, \dots, f_n) = (f_1^* a, f_2 \cdots, f_n)$$
$$+ \sum_{i=1}^{n-1} (-1)^i (a; f_1, \dots, f_i f_{i+1}, \dots, f_n) + (-1)^n (f_{n*} a, f_1, \dots, f_{n-1}).$$

The homology $\mathrm{H}_*(\mathcal{C}, D)$ of the category $\mathcal{C}$ with coefficients in the bifunctor $D$ is defined as the homology of the complex $\mathrm{F}_*(\mathcal{C}, D)$. Sometimes we write $\mathrm{H}^*(\mathcal{C}, (X, Y) \mapsto D(X, Y))$ to make explicit the values of the bifunctor $D$. The category of bifunctors $\mathrm{Func}(\mathcal{C}^{\mathrm{op}} \times \mathcal{C}, \mathrm{Ab})$ is a category of functors, hence it is an abelian category with enough projective and injective objects. Lemma 2.4 applied to the category $\mathcal{C}^{\mathrm{op}} \times \mathcal{C}$ says that projective generators are given by

$$P_{A,B} = \mathbb{Z}[\mathrm{Hom}_{\mathcal{C}}(-, A) \times \mathrm{Hom}_{\mathcal{C}}(B, -)], \quad A, B \in \mathcal{C}.$$

**Lemma 2.8**. — *For any $A$, $B$ in $\mathcal{C}$ one has:*

$$\mathrm{H}_n(\mathcal{C}, P_{A,B}) = 0 \quad \text{if} \quad n > 0, \quad \text{and} \quad \mathrm{H}_0(\mathcal{C}, P_{A,B}) = \mathbb{Z}[\mathrm{Hom}_{\mathcal{C}}(B, A)].$$

*Proof.* — Since $\mathrm{F}_0(\mathcal{C}, P_{A,B})$ is the free abelian group on the set of diagrams $A \leftarrow X \leftarrow B$, composition yields an homomorphism

$$\mathrm{F}_0(\mathcal{C}, P_{A,B}) \longrightarrow \mathbb{Z}[\mathrm{Hom}_{\mathcal{C}}(B, A)].$$

Let $\mathrm{F}_{-1}(\mathcal{C}, P_{A,B})$ be $\mathbb{Z}[\mathrm{Hom}_{\mathcal{C}}(B, A)]$. For $n \geqslant -1$, $\mathrm{F}_n(\mathcal{C}, P_{A,B})$ is the free abelian group spanned by the diagrams

$$A \xleftarrow{\ f\ } X_0 \xleftarrow{\ f_1\ } X_1 \xleftarrow{\ f_2\ } \cdots \xleftarrow{\ f_n\ } X_n \xleftarrow{\ g\ } B.$$

A contracting homotopy $h_n : \mathrm{F}_n(\mathcal{C}, P_{A,B}) \to \mathrm{F}_{n+1}(\mathcal{C}, P_{A,B})$, $n \geqslant -1$ is defined by

$$h_n(f, f_1, \ldots, f_n, g) = (\mathrm{Id}_A, f, f_1, \ldots, f_n, g). \qquad \square$$

**Corollary 2.9**. — *The sequence of functors*

$$\big(\mathrm{H}_n(\mathcal{C}, -) : \mathrm{Func}(\mathcal{C}^{\mathrm{op}} \times \mathcal{C}, \mathrm{Ab}) \longrightarrow \mathrm{Ab}\big)_{n \geqslant 0}$$

*is universal.*

*Proof.* — Since the functor $\mathrm{F}_n(\mathcal{C}, -) : \mathrm{Func}(\mathcal{C}^{\mathrm{op}} \times \mathcal{C}, \mathrm{Ab}) \to \mathrm{Ab}$ is exact, it follows that $\mathrm{H}_n(\mathcal{C}, -)$ is an exact connected sequence of functors. It is universal thanks to Lemma 2.8. $\qquad \square$

We now express the homology of small categories as Tor-groups. Let $R$ be a ring and let $F : \mathcal{C}^{\mathrm{op}} \to \mathrm{Mod}\text{-}R$ be a contravariant functor to the category of right $R$-modules. Let $T : \mathcal{C} \to R\text{-Mod}$ be a covariant functor to the category of left $R$-modules. Then

$$(X, Y) \longmapsto F(X) \otimes_R T(Y)$$

defines a bifunctor $T \boxtimes_R F : \mathcal{C}^{\mathrm{op}} \times \mathcal{C} \to \mathrm{Ab}$.

**Proposition 2.10**. — *Assume that the values of $F$ or $T$ are projective $R$-modules. Then, for each $i \geqslant 0$, there is an isomorphism*

$$\mathrm{Tor}_i^{\mathcal{C}}(F, T) \cong \mathrm{H}_i(\mathcal{C}, F \boxtimes_R T).$$

*Proof.* — Since

$$\mathrm{H}_0(\mathcal{C}, F \boxtimes_R T) = \mathrm{Coker}\left[\bigoplus_{f : X \to Y} F(Y) \otimes_R T(X) \longrightarrow \bigoplus_X F(X) \otimes_R T(X)\right] \cong F \otimes_{\mathcal{C}} T$$

we have the expected isomorphism for all $T$ and $F$ when $i = 0$. Assume now that the values of $F$ are projective. Varying $T$ we obtain an exact connected sequence of functors

$$\mathrm{H}_n(\mathcal{C}, F \boxtimes_R (-)) : \mathrm{Func}(\mathcal{C}, R\text{-Mod}) \longrightarrow \mathrm{Ab}, \ n \geqslant 0.$$

Thus it suffices to show $\mathrm{H}_n(\mathcal{C}, F \boxtimes_R P_A) = 0$ for $n > 0$, where as usual $P_A(Y) = R[\mathrm{Hom}_{\mathcal{C}}(A, Y)]$. Since

$$\mathrm{F}_0(\mathcal{C}, F \boxtimes_R P_A) \cong \bigoplus_{X \leftarrow A} F(X)$$

projection on the identity factor yields a map $\mathrm{F}_0(\mathcal{C}, F \boxtimes_R P_A) \to F(A)$. Let $\mathrm{F}_{-1}(\mathcal{C}, F \boxtimes_R P_A)$ be $F(A)$. For $n \geqslant -1$,

$$\mathrm{F}_n(\mathcal{C}, F \boxtimes_R P_A) = \bigoplus_{X_0 \leftarrow \cdots \leftarrow X_n \leftarrow A} F(X_0).$$

A contracting homotopy $h_n : \mathrm{F}_n(\mathcal{C}, F \boxtimes_R P_A) \to \mathrm{F}_{n+1}(\mathcal{C}, F \boxtimes_R P_A)$, $n \geqslant -1$ is defined by:

$$h_n(a; f_1, \ldots, f_n, g) \longmapsto (a; f_1, \ldots, f_n, g, \mathrm{Id}_A). \qquad \square$$

**Remark 2.11**. — Indeed the following more general result is true (compare with [**11**, Theorem B]): For any functors $F$ and $T$, there is a spectral sequence

$$E^2_{pq} = \mathrm{H}_p(\mathcal{C}, (X, Y) \longmapsto \mathrm{Tor}^R_q(FX, TY)) \Longrightarrow \mathrm{Tor}^{\mathcal{C}}_*(F, T).$$

This is a consequence of Grothendieck's spectral sequence for a composite of functors.

**Remark 2.12**. — Take $\mathcal{C}$ to be the category $\mathbb{P}(R)$ of finitely generated projective modules over a ring $R$. For any functor $T : \mathbb{P}(R) \to R\text{-Mod}$, the *MacLane homology* $\mathrm{HML}_*(R, T)$ of $R$ *with coefficient in* $T$ is $\mathrm{Tor}^{\mathbb{P}}_*(\mathrm{Id}^*, T)$, where $\mathrm{Id}^* = \mathrm{Hom}_R(-, R)$. Proposition 2.10 shows that $\mathrm{HML}_*(R, T) \cong \mathrm{H}_*(\mathbb{P}(R), D_T)$, where $D_T(X, Y) = \mathrm{Hom}_R(X, T(Y))$.

Another application of Proposition 2.10 is the following. Consider the constant functor with value $R$, still denoted by $R$. Note that

$$(R \boxtimes_R T)(X, Y) = T(Y)$$

is a bifunctor which is constant with respect to the contravariant argument. Let us denote this bifunctor again by $T$. Thus

$$\mathrm{H}_*(\mathcal{C}, T) \cong \mathrm{Tor}^{\mathcal{C}}_*(R, T).$$

Since $R \otimes_{\mathcal{C}} T = \mathrm{colim}\, T$, the sequence of functors $\mathrm{H}_*(\mathcal{C}, -) \colon \mathrm{Func}(\mathcal{C}, R\text{-Mod}) \to \mathrm{Ab}$ is isomorphic to the left derived functors of the functor

$$\mathrm{colim} : \mathrm{Func}(\mathcal{C}, R\text{-Mod}) \longrightarrow \mathrm{Ab}.$$

**Corollary 2.13**. — *If the category $\mathcal{C}$ has a terminal object $C$, then for any functor $T : \mathcal{C} \to R\text{-Mod}$: $\mathrm{H}_i(\mathcal{C}, T) = 0$ for positive $i$, and $\mathrm{H}_0(\mathcal{C}, T) = T(C)$.*

*Proof.* — It is clear that $\mathrm{colim}(T) = T(C)$. Thus colim is an exact functor and the result follows. $\qquad \square$

**Proposition 2.14**. — *Let $\mathcal{C}$ be a category, with finite coproducts and finite products. For any bifunctor $D : \mathcal{C}^{\mathrm{op}} \times \mathcal{C} \to \mathrm{Ab}$, let $D_{\coprod}$ and $D_{\prod}$ be the bifunctors defined on $\mathcal{C}$ by $D_{\coprod}(X, Y) = D(X \coprod X, Y)$ and $D_{\prod}(X, Y) = D(X, Y \times Y)$. There is an isomorphism*

$$\mathrm{H}_*(\mathcal{C}, D_{\coprod}) \cong \mathrm{H}_*(\mathcal{C}, D_{\prod}).$$

*Proof.* — By varying $D \in \mathrm{Func}(\mathcal{C}^{\mathrm{op}} \times \mathcal{C}, \mathrm{Ab})$ we obtain two exact connected sequences of functors $D \mapsto \mathrm{H}_n(\mathcal{C}, D_{\coprod})$, $n \geqslant 0$ and $D \mapsto \mathrm{H}_n(\mathcal{C}, D_{\prod})$, $n \geqslant 0$. It suffices to show that both of them are universal and take the same values on projective objects. Consider a projective $P_{A,B} \colon P_{A,B}(X, Y) = \mathbb{Z}[\mathrm{Hom}_{\mathcal{C}}(X, A) \times \mathrm{Hom}_{\mathcal{C}}(B, Y)]$. One finds:

$(P_{A,B})_{\amalg} \cong P_{A \times A, B}$ and $(P_{A,B})_{\prod} \cong P_{A, B \amalg B}$. It follows from Lemma 2.8 that homology vanishes in positive dimensions for both functors, and it equals

$$\mathbb{Z}[\mathrm{Hom}_{\mathcal{C}}(B, A \times A)] \cong \mathbb{Z}[\mathrm{Hom}_{\mathcal{C}}(B \coprod B, A)]$$

in dimension 0. Lemma 2.1 finishes the proof. $\square$

## 3. Finite degree functors

**3.1. Cross-effects.** — Let $F : \mathcal{C} \to \mathbf{A}$ be a functor from an additive category to an abelian category. For all $X$ and $Y$ in $\mathcal{C}$, the projections induce a natural map:

$$F(X \oplus Y) \longrightarrow F(X) \oplus F(Y).$$

Suppose $F(0) = 0$. This map is an epimorphism, which is naturally split by the map induced by the inclusions of $X$ and $Y$ in $X \oplus Y$. The *second cross-effect* of $F$ is the bifunctor defined by

$$(\mathrm{Cr}_2 F)(X, Y) := \mathrm{Ker}\left[F(X \oplus Y) \longrightarrow F(X) \oplus F(Y)\right].$$

Since the cross-effect fits in a natural splitting:

$$F(X \oplus Y) \cong F(X) \oplus F(Y) \oplus \mathrm{Cr}_2 F(X, Y),$$

the functor $F$ is additive if and only if $\mathrm{Cr}_2 F = 0$.

In order to define the third cross-effect of $F$, we proceed as follows. We consider the second cross-effect $(\mathrm{Cr}_2 F)(X, Y)$. We fix $Y$ and let $X$ vary. In this way we obtain the functor $X \mapsto (\mathrm{Cr}_2 F)(X, Y)$ which we take the second cross-effect of. We can continue this process and define the $n$-th cross-effect $(\mathrm{Cr}_n F)(X_1, \ldots, X_n)$. Alternatively, $\mathrm{Cr}_n F(X_1, \ldots, X_n)$ is isomorphic to the kernel of the natural homomorphism

$$F(X_1 \oplus \cdots \oplus X_n) \longrightarrow \bigoplus_{i=1}^{n} F(X_1 \oplus \cdots \oplus \widehat{X}_i \oplus \cdots \oplus X_n)$$

This shows that the $n$-th cross-effect $(\mathrm{Cr}_n F)(X_1, \ldots, X_n)$ is symmetric on $X_1, \ldots, X_n$. Another consequence of the definition is the following natural splitting:

$$(2) \qquad F(X_1 \oplus X_2 \oplus \cdots \oplus X_n) \cong \bigoplus_{k=1}^{n} \bigoplus_{1 \leqslant i_1 < \cdots < i_k \leqslant n} (\mathrm{Cr}_k F)(X_{i_1}, \ldots, X_{i_k})$$

One observes that an arbitrary functor $F : \mathcal{C} \to \mathbf{A}$ has a natural decomposition $F \cong F(0) \oplus F'$ with $F'(0) = 0$. This allows to define the $n$-th cross-effects of $F$ to be the cross-effects of $F'$ for any $n \geqslant 2$. It is be convenient to call the functor $F'$ the *first* cross-effect of $F$.

## 3.2. Functors of finite degree

**Definition 3.1** ([7]). — A functor $F: \mathcal{C} \to \mathbf{A}$ is of degree $n$ if its $(n+1)$-st cross-effect functor vanishes, but its $n$-th cross-effect does not. We then write: $\deg(F) = n$.

**Example 3.2**. — Assume that $\mathcal{C}$ and $\mathbf{A}$ are the category of (left) modules over a commutative ring $R$. The functors $X \mapsto \Lambda^n X$, $S^n X$, $X^{\otimes n}$ are of degree $n$, while the functor $X \mapsto R[X]$ is not of finite degree.

For an integer $d$, we let $\mathrm{Func}_d(\mathcal{C}, \mathbf{A})$ be the full subcategory of $\mathrm{Func}(\mathcal{C}, \mathbf{A})$ of functors of degree $\leqslant d$. By definition, $\mathrm{Func}_0(\mathcal{C}, \mathbf{A})$ consists of constant functors. The subcategory $\mathrm{Func}_d(\mathcal{C}, \mathbf{A})$ is closed in respect of coproducts and products. It is closed also in respect of subobjects, quotients and extensions. It follows that the category $\mathrm{Func}_d(\mathcal{C}, \mathbf{A})$ is an abelian category, and the inclusion $\mathrm{Func}_d(\mathcal{C}, \mathbf{A}) \subset \mathrm{Func}(\mathcal{C}, \mathbf{A})$ is an exact functor.

We now construct a left adjoint to this inclusion functor. Such a left adjoint can be seen as a Taylor expansion at order $d$. Consider the following "codiagonal" morphism:

$$(1_X, \ldots, 1_X) : X^{\oplus(d+1)} = X \oplus X \oplus \cdots \oplus X \longrightarrow X.$$

For a functor $F : \mathcal{C} \to \mathbf{A}$, the codiagonal induces a natural map $F(X \oplus \cdots \oplus X) \to F(X)$, whose restriction on $(\mathrm{Cr}_{d+1} F)(X, \ldots, X)$ defines a morphism

$$\varrho_{d,X}(F) : (\mathrm{Cr}_{d+1} F)(X, \ldots, X) \longrightarrow F(X).$$

Note (for use in Definition 4.2) that the above formula defines a natural transformation to $F$ which we denote $\varrho_d(F)$ (of course also natural in $F$). A left adjoint functor $t_d : \mathrm{Func}(\mathcal{C}, \mathbf{A}) \to \mathrm{Func}_d(\mathcal{C}, \mathbf{A})$ is given by:

$$(t_d F)(X) = \mathrm{Coker}\left(\varrho_{D,X}(F) : (\mathrm{Cr}_{d+1} F)(X, \ldots, X) \longrightarrow F(X)\right).$$

Similarly, a right adjoint functor $t^d : \mathrm{Func}(\mathcal{C}, \mathbf{A}) \to \mathrm{Func}_d(\mathcal{C}, \mathbf{A})$ is defined using the diagonal morphism $X \to X^{\oplus(d+1)}$. Since $(\mathrm{Cr}_{d+1}) F(X, \ldots, X)$ is a direct summand of $F(X^{d+1})$, we obtain a natural transformation

$$\vartheta_{D,X}(F) : F(X) \longrightarrow (\mathrm{Cr}_{d+1} F)(X, \ldots, X)$$

and a right adjoint is defined by: $(t^d F)(X) = \mathrm{Ker}(\vartheta_{D,X}(F))$. Observe that the natural transformation $F \to t_d F$ is an isomorphism if and only if $\deg(F) \leqslant d$. Similarly, $t^d F \to F$ is an isomorphism if and only if $\deg(F) \leqslant d$. Since $\mathrm{Func}(\mathcal{C}, R\text{-Mod})$ has enough projective and injective objects the same is also true for $\mathrm{Func}_d(\mathcal{C}, R\text{-Mod})$. We sum up this discussion with the following lemma.

**Lemma 3.3**. — *The inclusion functor* $\mathrm{Func}_d(\mathcal{C}, \mathbf{A}) \subset \mathrm{Func}(\mathcal{C}, \mathbf{A})$ *has a left adjoint (and a right adjoint). The category* $\mathrm{Func}_d(\mathcal{C}, R\text{-Mod})$ *is an abelian category with enough projectives (and enough injectives).*

**3.3. A cancellation lemma.** — For a bifunctor $D : \mathcal{C} \times \mathcal{C} \to \mathrm{Ab}$ we let $\deg_1 D$ (resp. $\deg_2 D$) be the degree with respect to the first (resp. second) variable. Similarly, $\mathrm{Cr}_n^1 D$ and $\mathrm{Cr}_n^2 D$ denote the $n$-th cross-effect functor with respect to the first and the second variable respectively. The following vanishing result is a variant of the vanishing result of the second author [**14**] (see Section 2.4 of *Introduction to functor homology* in this volume). The lemma states that, after taking homology, cross-effects can be moved from one argument of the bifunctor coefficients to the other.

**Lemma 3.4**. — *Let $\mathcal{C}$ be an additive category and let $D : \mathcal{C}^{\mathrm{op}} \times \mathcal{C} \to \mathrm{Ab}$ be a bifunctor. There is a natural isomorphism:*

$$\mathrm{H}_*(\mathcal{C}, (X,Y) \longmapsto (\mathrm{Cr}_n^1)D(X,\ldots,X;Y)) \cong \mathrm{H}_*(\mathcal{C}, (X,Y) \longmapsto (\mathrm{Cr}_n^2 D)(X;Y,\ldots,Y)).$$

*In particular, if $\deg_1 D < n$, then:*

$$\mathrm{H}_*(\mathcal{C}, (X,Y) \longmapsto (\mathrm{Cr}_n^2 D)(X;Y,\ldots,Y)) = 0.$$

*Proof.* — By a slight generalization of Lemma 2.14 to $n+1$ factors:

$$\mathrm{H}_*\big(\mathcal{C}, (X,Y) \mapsto D(X \oplus \cdots \oplus X;Y)\big) \cong \mathrm{H}_*\big(\mathcal{C}, (X,Y) \mapsto D(X;Y \oplus \cdots \oplus Y)\big).$$

Consideration of the splitting (2) in 3.1 allows to deduce the first isomorphism. If $n > \deg_1 D$, the left hand side is zero, thus the same is true for the right hand side. $\quad\square$

## 4. Proof of Scorichenko's Theorem

Scorichenko's proof uses an intermediary category, the category $\mathbb{E}$, which still has finitely generated projective $R$-modules as objects, but where morphisms are epimorphisms of $R$-modules. Scorichenko observed that the definition of stable $K$-theory is still meaningful for bifunctors from $\mathbb{E}^{\mathrm{op}} \times \mathbb{P}$. This leads to Theorem 4.1 computing stable K-theory as homology of $\mathbb{E}$, for those bifunctors having finite degree in the covariant argument. Theorem 4.4 then compares the homology of $\mathbb{E}$ and the homology of $\mathbb{P}$, this time for bifunctors having finite degree in the contravariant argument. The two comparisons together give Theorem 0.1.

**4.1. Stable $K$-theory is homology of the category $\mathbb{E}$.** — The first main step is similar in spirit to the approach given in [**4**], where stable $K$-theory was obtained as the derived functor of the functor

$$K_0^{\mathrm{st}} : \mathrm{Func}(\mathbb{P}^{\mathrm{op}}(R) \times \mathbb{P}(R), \mathrm{Ab}) \longrightarrow \mathrm{Ab}$$

but under restricted conditions on the ring $R$ (see Remark 5.3). Extending the domain of definition of stable $K$-theory to the category $\mathrm{Func}(\mathbb{E}^{\mathrm{op}} \times \mathbb{P}, \mathrm{Ab})$ overcomes this difficulty.

We let $\mu$ be the inclusion $\mathbb{E} \subset \mathbb{P}$. Thus $\mu$ is the identity on objects.

***Theorem 4.1***. — *Let $D : \mathbb{E}^{\mathrm{op}}(R) \times \mathbb{P}(R) \to \mathrm{Ab}$ be a bifunctor, and let $\mu^* D$ be the composite:*

$$\mathbb{E}(R)^{\mathrm{op}} \times \mathbb{E}(R) \xrightarrow{\ \mathrm{Id} \times \mu\ } \mathbb{E}(R)^{\mathrm{op}} \times \mathbb{P}(R) \xrightarrow{\ D\ } \mathrm{Ab}$$

*If $D$ has finite degree with respect to the covariant argument, then*

$$K^{\mathrm{st}}_*(R, D) \cong \mathrm{H}_*(\mathbb{E}(R), \mu^*(D)).$$

*Proof.* — First we consider the case when $D$ is constant with respect to the first variable. In this case, the right hand side is the homology of the category $\mathbb{E}$ with coefficients in a functor $T = D(0, -)$. Because $0$ is the terminal object of $\mathbb{E}$, this homology vanishes in positive dimensions and it equals $T(0)$ in dimension $0$ thanks to Corollary 2.13. The corresponding statement for stable $K$-theory is a result of Betley [**2**].

Next, we extend the comparison to bifunctors $Z_{U,B}$ defined by

$$Z_{U,B} = B(Y)[\mathrm{Hom}_{\mathbb{E}}(X, U)] = P_U(X) \otimes B(Y),$$

where $B : \mathbb{P} \to \mathrm{Ab}$ is a finite degree functor, and $U \in Ob(\mathbb{E}) = Ob(\mathbb{P})$. Lemma 2.10 tells us that the right-hand side $\mathrm{H}_*(\mathbb{E}, \mu^* D)$ is isomorphic to $\mathrm{Tor}^{\mathbb{E}}_*(P_U, \mu^* B)$. Since $P_U$ is a projective in $\mathrm{Func}(\mathbb{E}, \mathrm{Ab})$, these groups vanish in positive dimensions and $\mathrm{Tor}^{\mathbb{E}}_0(P_U, \mu^* B)$ is isomorphic to $B(U)$.

Let us show the same result for $K^{\mathrm{st}}_*(R, Z_{U,B})$. Having fixed a projection $\pi : R^\infty \to U$ and the corresponding stabilizer $\mathrm{Stab}(\pi)$, we get an isomorphism of $\mathrm{GL}(R)$-modules

$$Z^\infty_{U,B} \cong \mathrm{Ind}^{\mathrm{GL}(R)}_{\mathrm{Stab}(\pi)} B^\infty.$$

It follows from the Shapiro lemma in group homology that

$$\mathrm{H}_*(\mathrm{GL}(R), Z^\infty_{U,B}) \cong \mathrm{H}_*(\mathrm{Stab}(\pi), B^\infty).$$

It is known that $\mathrm{H}_*(\mathrm{Stab}(\pi), B^\infty) \cong \mathrm{H}_*(\mathrm{GL}(R), B(U \oplus \infty))$. Because $B$ has finite degree, we can still use the result of Betley [**2**] to conclude that $\mathrm{H}_*(\mathrm{GL}(R), Z^\infty_{U,B}) \cong \mathrm{H}_*(\mathrm{GL}(R), B(U))$ where $\mathrm{GL}(R)$ acts trivially on $B(U)$. Now the spectral sequence (1) yields that $K^{\mathrm{st}}$ vanishes on $Z_{U,B}$ in positive dimensions and it equals $B(U)$ in dimension $0$.

To conclude, consider, for each integer $d$, the abelian subcategory of $\mathrm{Func}(\mathbb{E} \times \mathbb{P}, \mathrm{Ab})$ consisting of bifunctors $D : \mathbb{E}(R) \times \mathbb{P}(R) \to \mathrm{Ab}$ which have degree $d$ with respect to the first variable. Both terms in the statement form an exact connected sequence of functors defined on this subcategory. The result follows from Proposition 2.1. $\square$

**4.2. Another cancellation lemma.** — As pointed out in Example 3.2, while most of the usual functors have finite degree, the cross-effect on projective functors has the opposite property of enlarging their size. Scorichenko applies Lemma 3.4 to bifunctors which are of finite degree in one variable, but have this opposite property with respect to the other variable. Lemma 4.3's homology cancellation was known to the first author in the special case when $R$ is a prime field and when $D$ is the bifunctor defined by: $D(X, Y) = \mathrm{Hom}(F(X), P_A(Y))$ for a finite degree functor $F$ and a projective $P_A$ (for a nice proof due to L. Schwartz, see [**17**, Appendix]).

***Definition 4.2***. — Let $D : \mathbb{P}^{\mathrm{op}} \times \mathbb{P} \to \mathrm{Ab}$ be a bifunctor and let $d$ be an integer. Consider the natural transformation $\varrho_d(D)$:

$$\varrho_{d,X,Y}(D) : (\mathrm{Cr}_{d+1}^2 D)(X; Y, \ldots, Y) \longrightarrow D(X, Y).$$

The bifunctor $D$ is called $S_d$-acyclic if for each $X, Y \in \mathbb{P}$ the morphism $\varrho_{d,X,Y}$ has a section

$$s_{X,Y} : D(X, Y) \longrightarrow (\mathrm{Cr}_{d+1}^2 D)(X; Y, \ldots, Y)$$

which is natural on $X \in \mathbb{P}$ and on $Y \in \mathbb{E}$.

***Lemma 4.3***. — *Let $D : \mathbb{P}^{\mathrm{op}} \times \mathbb{P} \to \mathrm{Ab}$ be a bifunctor and assume that $D$ is of degree $\leqslant d$ with respect to the first variable. If $D$ is $S_d$-acyclic then:* $\mathrm{H}_*(\mathbb{P}, D) = 0$.

*Proof*. — By assumption the natural transformation $\varrho_d$ is surjective on $D$. We let $C$ be the kernel of this transformation. Direct computation checks that if $D$ is $S_d$-acyclic then $C$ is $S_d$-acyclic as well. It follows from Lemma 3.4 that $\mathrm{H}_0(\mathbb{P}, D) = 0$ and $\mathrm{H}_{i+1}(\mathbb{P}, D) \cong \mathrm{H}_i(\mathbb{P}, C)$. Now one can use induction to finish the proof. □

**4.3. The homology of the category $\mathbb{E}$.** — We now turn to Scorichenko's second theorem.

***Theorem 4.4***. — *Let $D : \mathbb{P}^{\mathrm{op}} \times \mathbb{P} \to \mathrm{Ab}$ be a bifunctor. If $D$ has finite degree with respect to the contravariant argument, then the inclusion $\mu : \mathbb{E} \subset \mathbb{P}$ yields an isomorphism in homology*

$$\mathrm{H}_*(\mathbb{E}, \mu^* D) \cong \mathrm{H}_*(\mathbb{P}, D),$$

*where $\mu^* D : \mathbb{E}^{\mathrm{op}} \times \mathbb{E} \to \mathrm{Ab}$ is the composite*

$$\mathbb{E}^{\mathrm{op}} \times \mathbb{E} \xrightarrow{\mu \times \mu} \mathbb{P}^{\mathrm{op}} \times \mathbb{P} \xrightarrow{D} \mathrm{Ab}.$$

We refer to the original paper [**20**] for the proof of this theorem in its generality. Here we give the proof in the case, when submodules of finitely generated projective left $R$-modules are still finitely generated and projective. This holds for instance when $R$ is a Dedekind domain, and more generally when $R$ is left noetherian and $\mathrm{gl.dim}(R) \leqslant 1$. In the rest of Section 4 we shall assume that $R$ satisfies this condition.

Let $\mu : \mathbb{E} \to \mathbb{P}$ be the inclusion. As does any functor between small categories, it yields a functor

$$\mu^* : \mathrm{Func}(\mathbb{P}, \mathrm{Ab}) \longrightarrow \mathrm{Func}(\mathbb{E}, \mathrm{Ab}), \quad T \longmapsto T \circ \mu,$$

which has both left and right adjoint functors known as *left* and *right Kan extensions of the functor* $\mu$ [**13**]. We shall need only the left Kan extension $\mu_!$. The hypothesis on the ring allows canonical factorisation of linear maps by epimorphisms, and thus allows the following description of the left Kan extension of $\mu$. For a functor $T : \mathbb{E} \to \mathrm{Ab}$ and a finitely generated projective left $R$-module $P$:

$$\mu_! T(P) := \bigoplus_{W \subset P} T(W)$$

where $W$ runs through the submodules of $P$. The hypothesis on the ring $R$ insures that any such $W$ is also finitely generated and projective, and $T(W)$ is thus well-defined. A typical generator of $\mu_! T(P)$ is denoted by $(a; W)$, where $a \in T(W)$. If $Q$ is an another finitely generated projective $R$-module and $f : P \to Q$ is $R$-linear, one defines $\mu_!(f) : \mu_! T(P) \to \mu_! T(Q)$ by

$$\mu_!(f)(a, W) = (T(f')(a), f(W)),$$

where $f' : W \to f(W)$ is the restriction of $f$. We obtain a functor

$$\mu_! : \mathrm{Func}(\mathbb{E}, \mathrm{Ab}) \longrightarrow \mathrm{Func}(\mathbb{P}, \mathrm{Ab})$$

which is left adjoint to $\mu^*$. The Kan extension $\mu_!$ is the functor defined by Suslin in the case when $R$ is a finite field [**8**, Appendix] (it is denoted $\widetilde{a}$ there).

This construction bears some obvious variations. For example we also have the functor

$$\mu^* : \mathrm{Func}(\mathbb{P}^{\mathrm{op}} \times \mathbb{E}, \mathrm{Ab}) \longrightarrow \mathrm{Func}(\mathbb{E}^{\mathrm{op}} \times \mathbb{E}, \mathrm{Ab}), \quad \mu^* B = B \circ (\mu \times \mathrm{Id}_{\mathbb{E}})$$

and the functor

$$\mu_! : \mathrm{Func}(\mathbb{P}^{\mathrm{op}} \times \mathbb{E}, \mathrm{Ab}) \longrightarrow \mathrm{Func}(\mathbb{P}^{\mathrm{op}} \times \mathbb{P}, \mathrm{Ab})$$

which is given by: $(\mu_! B)(X, Y) = \bigoplus_{W \subset Y} B(X, W)$.

**Lemma 4.5**. — *For each $A$ and $B$, the equation*

$$S_{A,B}(X, Y) = \mathbb{Z}[\mathrm{Hom}_{\mathbb{P}}(X, A) \times \mathrm{Hom}_{\mathbb{E}}(B, Y)]$$

*defines an object $S_{A,B}$ of the category $\mathrm{Func}(\mathbb{P}^{\mathrm{op}} \times \mathbb{E}, \mathrm{Ab})$. These objects are projective generators of the category $\mathrm{Func}(\mathbb{P}^{\mathrm{op}} \times \mathbb{E}, \mathrm{Ab})$. Furthermore, the following isomorphisms hold:*

$$\mathrm{H}_i(\mathbb{P}, \mu_! S_{A,B}) = 0 = \mathrm{H}_i(\mathbb{E}, \mu^* S_{A,B}), \quad if \ i > 0,$$

*and*

$$\mathrm{H}_0(\mathbb{P}, \mu_! S_{A,B}) \cong \mathbb{Z}[\mathrm{Hom}_{\mathbb{P}}(B, A)] \cong \mathrm{H}_0(\mathbb{E}, \mu^* S_{A,B}).$$

*Proof*. — The statement on projective generators follows from Lemma 2.4. We have the following bijection

$$(3) \qquad\qquad \mathrm{Hom}_{\mathbb{P}}(X, A) \cong \coprod_{W \subset A} \mathrm{Hom}_{\mathbb{E}}(X, W)$$

and it is natural on $(X, A) \in \mathbb{E} \times \mathbb{P}$. It follows that

$$\mu_!(S_{A,B})(X, Y) = \bigoplus_{U \subset Y} \mathbb{Z}[\mathrm{Hom}_{\mathbb{P}}(X, A) \times \mathrm{Hom}_{\mathbb{E}}(B, U)]$$
$$\cong \mathbb{Z}[\mathrm{Hom}_{\mathbb{P}}(X, A) \times \mathrm{Hom}_{\mathbb{P}}(B, Y)].$$

Thus $\mu_!(S_{A,B})$ is a standard projective generator of $\mathrm{Func}(\mathbb{P}^{\mathrm{op}} \times \mathbb{P}, \mathrm{Ab})$ and therefore Lemma 2.8 shows that

$$\mathrm{H}_i(\mathbb{P}, \mu_! S_{A,B}) = 0, \text{ if } i > 0, \quad \text{and } \mathrm{H}_0(\mathbb{P}, \mu_! S_{A,B}) = \mathbb{Z}[\mathrm{Hom}_{\mathbb{P}}(B, A)]$$

Similarly, we have

$$\mu^*(S_{A,B})(X, Y) = \mathbb{Z}[\mathrm{Hom}_{\mathbb{P}}(X, A) \times \mathrm{Hom}_{\mathbb{E}}(B, Y)]$$
$$\cong \bigoplus_{W \subset A} \mathbb{Z}[\mathrm{Hom}_{\mathbb{E}}(X, W) \times \mathrm{Hom}_{\mathbb{E}}(B, Y)].$$

Thus: $\mu^*(S_{A,B}) \cong \bigoplus_{W \subset B} P_{W,B}$ , and it is projective in $\mathrm{Func}(\mathbb{E}^{\mathrm{op}} \times \mathbb{E}, \mathrm{Ab})$. Lemma 2.8 still applies to get: $\mathrm{H}_i(\mathbb{E}, \mu^* S_{A,B}) = 0$, if $i > 0$ and

$$\mathrm{H}_0(\mathbb{E}, \mu^* S_{A,B}) \cong \bigoplus_{W \subset B} \mathbb{Z}[\mathrm{Hom}_{\mathbb{E}}(W, A)] \cong \mathbb{Z}[\mathrm{Hom}_{\mathbb{P}}(B, A)]. \qquad \square$$

**Lemma 4.6**. — *Let $B : \mathbb{P}^{\mathrm{op}} \times \mathbb{E} \to \mathrm{Ab}$ be a bifunctor. For the bifunctors*

$$\mu_! B : \mathbb{P}^{\mathrm{op}} \times \mathbb{P} \longrightarrow \mathrm{Ab} \quad \text{and} \quad \mu^* B : \mathbb{E}^{\mathrm{op}} \times \mathbb{E} \longrightarrow \mathrm{Ab}$$

*there is an isomorphism*

$$\mathrm{H}_*(\mathbb{E}, \mu^* B) \cong \mathrm{H}_*(\mathbb{P}, \mu_! B).$$

*Proof*. — The sequences of functors

$$\mathrm{H}_n(\mathbb{E}, \mu^*(-)) : \mathrm{Func}(\mathbb{P}^{\mathrm{op}} \times \mathbb{E}, \mathrm{Ab}) \longrightarrow \mathrm{Ab}$$

and

$$\mathrm{H}_n(\mathbb{P}, \mu^*(-)) : \mathrm{Func}(\mathbb{P}^{\mathrm{op}} \times \mathbb{E}, \mathrm{Ab}) \longrightarrow \mathrm{Ab}$$

are exact connected sequences of functors. Lemma 4.5 shows that both of them are universal. Moreover, in dimension 0, both of them are isomorphic on projective generators. Since $\mathrm{H}_0$ is right exact and commutes with coproducts, it follows that both exact connected sequences of functors are isomorphic. $\qquad \square$

**Lemma 4.7**. — *For any bifunctor $D : \mathbb{P}^{\mathrm{op}} \times \mathbb{P} \to \mathrm{Ab}$ the natural map $\varepsilon : \mu_! \mu^* D \to D$ is an epimorphism and $\mathrm{Ker}(\varepsilon)$ is $S_d$-acyclic for any $d \geqslant 0$.*

*Proof*. — By definition: $(\mu_! \mu^* D)(X, Y) = \oplus_{W \subset Y} D(X, W)$. Let $(a, W)$, $a \in D(X, W)$, be a typical generator. The co-unit $\varepsilon$ sends $(a, W)$ to $i_{W*}(a)$, where $i_W : W \to Y$ is the inclusion. The first statement is clear by considering the factor for $W = Y$. We now prove the $S_d$-acyclicity of $\mathrm{Ker}(\varepsilon)$. Let us fix $d \geqslant 0$ and let us consider the map

$$s_{X,Y} : \mu_! \mu^* D(X, Y) = \bigoplus_{W \subset Y} D(X, W) \longrightarrow \mu_! \mu^* D(X, Y^{d+1}) = \bigoplus_{V \subset Y^{d+1}} D(X, V)$$

defined by:

$$s_{X,Y}(a, W) = (-j_* a, W \oplus Y^d) + (a, W).$$

Here $j : W \to W \oplus Y^d$ is given by $j(w) = (w, 0)$, while $W \oplus Y^d$ and $W$ are considered as submodules of $Y^{d+1}$ by embedding $W$ in the first factor. Thus $s_{X,Y}$ is natural on

$\mathbb{P}^{\mathrm{op}} \times \mathbb{E}$, and the image of $s$ lies in the subfunctor $\mathrm{Cr}_{d+1}^2 D$. Furthermore $\varrho \circ s(a, W) = -(i_{W*}(a), Y) + (a, W)$. Therefore the restriction of $\varrho \circ s$ on $\mathrm{Ker}(\varepsilon)$ is the identity. □

*Proof of Theorem 4.4.* — Let $\deg_1 D = d$. Since $\mathrm{Ker}(\varepsilon)$ is $S_d$-acyclic we have:

$$\mathrm{H}_*(\mathbb{P}, \mathrm{Ker}(\varepsilon)) = 0$$

thanks to Lemma 4.3. It follows from the exact sequence:

$$0 \longrightarrow \mathrm{Ker}(\varepsilon) \longrightarrow \mu_! \mu^* D \longrightarrow D \longrightarrow 0$$

that: $\mathrm{H}_*(\mathbb{P}, D) \cong \mathrm{H}_*(\mathbb{P}, \mu_! \mu^* D)$. We use Lemma 4.6 to finish the proof. □

## 5. General Linear homology with twisted coefficients

Let us recall from Section 1 the spectral sequence (1) for a ring $R$:

$$E_{pq}^2 = \mathrm{H}_p(\mathrm{GL}(R), K_q^{\mathrm{st}}(R, D)) \Longrightarrow \mathrm{H}_{p+q}(\mathrm{GL}(R), D_\infty).$$

In this section we show that the spectral sequence (1) degenerates at $E^2$, provided that the bifunctor $D$ takes vector space values and has finite degree with respect to the second argument. It is clear that the spectral sequence (1) is a functor of $D$.

We begin by showing that it is a $\partial$-spectral sequence. To define the maps $\partial$, we first extend the definition of stable K-theory to chain complexes of bifunctors. Since stable K-theory is defined by homology with twisted coefficients, it suffices to consider this case. Let $\Lambda$ be a ring, let $C_*$ be a complex of $\Lambda$-modules and let $M$ be a $\Lambda$-module, and consider the twisted homology $\mathrm{H}_*(C_* \otimes_\Lambda M)$. One can replace $M$ by a complex of modules and take the homology of the resulting total complex. When the complex $C_*$ is free, the resulting homology sends weak equivalences to isomorphisms. This situation occurs for the twisted homology of spaces, hence for stable K-theory. With such a definition, the spectral sequence (1) is now natural in respect of morphisms of chain complexes of bifunctors.

Let $0 \to A \to B \to C \to 0$ be a short exact sequence of bifunctors. It gives rise to a morphism from the complex

$$\cdots \longrightarrow 0 \longrightarrow 0 \longrightarrow C$$

to the complex

$$\cdots \longrightarrow 0 \longrightarrow B \longrightarrow C.$$

The latter is weakly equivalent to the complex $\cdots \to 0 \to A \to 0$. The induced map on spectral sequences yields the map $\partial$.

We now prove that the spectral sequence (1) satisfies the conditions of 2.2. Let $\mathrm{Func}_d$ be the category of bifunctors $D : \mathbb{P}^{\mathrm{op}} \times \mathbb{P} \to \mathrm{Ab}$ such that $\deg_2 D \leqslant d$. Let

$$0 \longrightarrow A \longrightarrow P \longrightarrow C \longrightarrow 0$$

be a short exact sequence in $\mathrm{Func}_d$ such that $P$ is a projective in $\mathrm{Func}_d$. It follows from Theorem 4.1 that the connecting homomorphism $K_q^{\mathrm{st}}(R, C) \to K_{q-1}^{\mathrm{st}}(R, A)$ is an isomorphism for $q \geqslant 1$ and is a monomorphism for $q = 1$. This applies also when considering only those functors taking values in vector spaces over a field. When $A$ takes vector space values, the monomorphism $K_1^{\mathrm{st}}(R, C) \to K_0^{\mathrm{st}}(R, A)$ splits. The same is true for the homomorphism $E_{pq}^2(C) \to E_{p,q-1}^2(A)$, because $\mathrm{GL}(R)$ acts trivially. We have proved:

**Theorem 5.1**. — *Let $D$ be a vector space values bifunctor which has finite degree with respect to the second argument. The spectral sequence (1)*

$$E_{pq}^2 = \mathrm{H}_p(\mathrm{GL}(R), K_q^{\mathrm{st}}(R, D)) \implies \mathrm{H}_{p+q}(\mathrm{GL}(R), D_\infty)$$

*stops at the $E_2$-term.*

**Remark 5.2**. — If the ring $R$ is such that

$$K_*^{\mathrm{st}}(R, -) : \mathrm{Func}(\mathbb{P}^{\mathrm{op}} \times \mathbb{P}, \mathrm{Ab}) \longrightarrow \mathrm{Ab}$$

is a universal connected sequence of functors, then Theorem 5.1 is true for any bifunctor $D : \mathbb{P}^{\mathrm{op}} \times \mathbb{P} \to \mathrm{Ab}$. This follows again from Lemma 2.2. Indeed, take $\mathbf{A} = \mathrm{Func}(\mathbb{P}^{\mathrm{op}} \times \mathbb{P}, \mathrm{Ab})$, and for a given $C \in \mathrm{Func}(\mathbb{P}^{\mathrm{op}} \times \mathbb{P}, \mathrm{Ab})$ take a standard projective $B$ as in [**4**, Lemma 2.1]. Then the connecting homomorphism $K_q^{\mathrm{st}}(R, C) \to K_{q-1}^{\mathrm{st}}(R, A)$ is an isomorphism for $q \geqslant 1$ and it is a monomorphism for $q = 1$. Because $K_0^{\mathrm{st}}(R, B)$ is a free abelian group, the monomorphism $K_1^{\mathrm{st}}(R, C) \to K_0^{\mathrm{st}}(R, A)$ splits. The above argument then applies. Indeed not only does the spectral sequence (1) stop at $E^2$, but no extension problem appears: there is a non-natural isomorphism $\mathrm{H}_n(\mathrm{GL}(R), D_\infty) \cong \bigoplus_{p+q=n} \mathrm{H}_p(\mathrm{GL}(R), K_q(R, D))$ [**4**].

**Remark 5.3**. — According to [**4**]

$$K_*^{\mathrm{st}}(R, -) : \mathrm{Func}(\mathbb{P}^{\mathrm{op}} \times \mathbb{P}, \mathrm{Ab}) \longrightarrow \mathrm{Ab}$$

is a universal connected sequence of functors provided that the ring $R$ is semi-simple. It was claimed in [**4**] that it is still the case for any commutative integral domain of finite Krull dimension, but the proof given there is not correct. However, because Lemma 1.6 of [**4**] is true under the condition that $R$ is a principal ideal domain, the statement holds for these rings.

**Example 5.4**. — Take $R = \mathbb{Z}$, and $D(X, Y) = \mathrm{Hom}(X, Y \otimes \mathbb{Z}/2\mathbb{Z})$. By [**9**, paragraphe 9.2], $\mathrm{H}_i(\mathbb{P}(\mathbb{Z}), D)$ is $\mathbb{Z}/2\mathbb{Z}$ when $i \equiv 0, 3 \pmod 4$ and 0 else. Scorichenko's theorem says it is the answer for $K_*^{\mathrm{st}}(\mathbb{Z}, D)$ as well. Recently [**1**] the Hopf algebra $\mathrm{H}_*(\mathrm{GL}(\mathbb{Z}), \mathbb{Z}/2\mathbb{Z})$ was computed; its Poincaré series is given by: $\prod_{n \geqslant 1} \frac{1 - t^{2n+1}}{1 - t^n}$. Theorem 5.1 implies that the Poincaré series of $\mathrm{H}_*(\mathrm{GL}(\mathbb{Z}), \mathrm{gl}(\mathbb{Z}/2\mathbb{Z}))$ is:

$$\frac{1 + t^3}{1 - t^4} \prod_{n \geqslant 1} \frac{1 - t^{2n+1}}{1 - t^n}.$$

# References

[1]  D. ARLETTAZ, M. MAMORU, N. KOJI & Y. NOBUAKI – The mod 2 cohomology of the linear groups over the ring of integers, *Proc. Amer. Math. Soc.* **127** (1999), p. 2199–2212.

[2]  S. BETLEY – Homology of $Gl(R)$ with coefficients in a functor of finite degree, *J. Algebra* **150** (1992), p. 73–86.

[3]  ———, Stable K-theory for finite fields, *K-Theory* **17** (1999), p. 103–111.

[4]  S. BETLEY & T. PIRASHVILI – Stable $K$-theory as a derived functor, *J. Pure Appl. Algebra* **96** (1994), p. 245–258.

[5]  H. CARTAN & S. EILENBERG – *Homological Algebra*, Princeton University Press, Princeton, NJ, 1956.

[6]  B.I. DUNDAS & R. MCCARTHY – Stable K-theory and topological Hochschild homology, *Ann. of Math.* **140** (1994), no. 3, p. 685–701, Erratum: *Ibid.* **142** (1995), no. 2, p. 425–426.

[7]  S. EILENBERG & S. MACLANE – On the groups $H(\pi, n)$ II, *Ann. of Math.* **60** (1954), p. 49–139.

[8]  V. FRANJOU, E. FRIEDLANDER, A. SCORICHENKO & A. SUSLIN – General linear and functor cohomology over finite fields, *Ann. of Math. (2)* **150** (1999), p. 663–728.

[9]  V. FRANJOU, J. LANNES & L. SCHWARTZ – Autour de la cohomologie de MacLane des corps finis, *Invent. Math.* **115** (1994), no. 3, p. 513–538.

[10] V. FRANJOU & T. PIRASHVILI – On the MacLane cohomology for the ring of integers, *Topology* **37** (1998), no. 1, p. 109–114.

[11] M. JIBLADZE & T. PIRASHVILI – Cohomology of algebraic theories, *J. Algebra* **137** (1991), p. 253–296.

[12] C. KASSEL – La K-théorie stable, *Bull. Soc. math. France* **110** (1982), p. 381–416.

[13] S. MACLANE – *Categories for the Working Mathematician*, Graduate Texts in Math., vol. 5, Springer-Verlag, New York-Berlin, 1971.

[14] T. PIRASHVILI – Higher additivizations, *Trudy Tbiliss. Mat. Inst. Razmadze Akad. Nauk Gruzin. SSR* **91** (1988), p. 44–54, Russian.

[15] ———, New homology and cohomology for rings, *Bull. Ac. Sc. Georgian SSR* **133** (1989), p. 477–480.

[16] T. PIRASHVILI & F. WALDHAUSEN – MacLane homology and topological Hochschild homology, *J. Pure Appl. Algebra* **82** (1992), no. 1, p. 81–98.

[17] C. POWELL – The Artinian conjecture for $I \otimes I$, *J. Pure Appl. Algebra* **128** (1998), p. 291–310.

[18] D. QUILLEN – Characteristic classes of representations, in *Algebraic K-theory (Proc. Conf., Northwestern Univ., Evanston, Ill., 1976)*, Lect. Notes in Math., vol. 551, Springer, Berlin, 1976, p. 189–216.

[19] R. SCHWÄNZL, R. STAFFELDT & F. WALDHAUSEN – Stable $K$-theory and topological Hochschild homology of $A_\infty$ rings, *Contemp. Math.* **199** (1996), p. 161–173.

[20] A. SCORICHENKO – Stable K-Theory and Functor Homology over a Ring, Thesis, Evanston, 2000.

[21] F. WALDHAUSEN – *Algebraic K-theory of topological spaces, II*, Lect. Notes in Math., vol. 763, Springer, 1979.

# INDEX OF NOTATION

**Categories**

$\mathbb{E}$, 119
$\mathcal{F}(\mathbb{K})$, 10
$\mathcal{F}$, 92
$\mathcal{F}_d$, 13
$\mathcal{F}_\omega$, 13
$\mathrm{Func}(\mathcal{C}, \mathcal{E})$, 9
$\mathrm{Func}(\mathcal{C}, \mathbf{A})$, 111
$\Gamma^d \mathcal{V}^f$, 21
$\mathcal{K}$, 105
$\mathcal{N}il$, 91
$\mathcal{P}$, 21, 41
$\mathcal{P}_d$, 21, 41
$\mathcal{P}ol_{n,d}$, 40
$\mathbb{P}(R)$, 108
$\mathfrak{S}_d\text{-Rep}$, 22
$\mathcal{U}$, 64, 104
$\boldsymbol{\mathcal{V}}$, 2
$\mathcal{V}^f$, 10

**Functors**

$D_\infty$, 109
$\gamma^d$, 21
$\Gamma^{d,m}$, 22
$\Gamma^n$, 4
$\mathrm{Id}$, 10
$I_V$, 11
$\Lambda^n$, 4
$p_n$, 98
$\mathrm{P}_V$, 11
$P_X$, 112
$R[S]$, 112

$\mathrm{S}^{d,m}$, 23
$\mathrm{S}^n$, 3
$\mathrm{T}^n$, 3
$\mathcal{A}_p^*$, 60
$\mathcal{A}_p$, 58, 60

$\mathrm{C}^{-1}$, 7
$\mathrm{Cr}_n^i D$, 119

$\deg_i D$, 119
$\Delta F$, 12

$\mathrm{Ext}_{\mathrm{Gen}}^i(F, T)$, 23

$F'(2)$, 63
$F(1)$, 62
$F^\sharp$, 11
$F(n)$, 65, 66

$\Gamma(L)$, 57
$\mathbb{G}_a$, 28
$\mathrm{GL}_{n(r)}$, 29

$\mathrm{HML}_*(R, T)$, 116
$\mathrm{H}_*(\mathcal{C}, D)$, 114
$\mathrm{H}^*(T)$, 17

$\mathrm{Ind}_H^G N$, 31

$J_*^*$, 77
$J(n)$, 76

$K_*^*$, 80
$k$, xiv

$\kappa_m$, 6
$kG$, xviii, 29
$k[G]$, 28
$K(i)$, 79
$K_n^i$, 19
$\mathrm{Kos}_*(f)$, 5
$\mathbb{K}[S]$, 2
$K_*^{\mathrm{st}}(R, D)$, 109

$M^\#$, 2
$M^G$, 2
$M_G$, 2

$N$, 5

$\Phi$, 2
$\mathrm{P}^i$, 61
$P(S, R)$, 61
$\mathbf{P}$, 61

$\mathfrak{S}_{i,j}$, 3
$\mathrm{Sq}^I$, 69
$\mathrm{Sq}^i$, 58
$\mathbf{Sq}$, 58

$t_E$, 87
$- \otimes_{\mathcal{C}} -$, 112
$\mathrm{Tor}_*^{\mathcal{C}}(-, -)$, 113
$T_V$, 82

$V^{(1)}$, 2, 32
$V^\#$, 2
$V^{(r)}$, 2

# INDEX

1-parameter subgroup, 49
$\partial$-spectral sequence, 110

abelian $p$-point, 51
Adams-Gunawardena-Miller theorem, 90
Adem relations, 67, 68
adjoint functors, 8
admissible monomial, 69
admissible sequence, 69
affine group, 60
affine group scheme, 28
analytic functor, 91

Boolean algebra, 90
Brown-Gitler modules, 74

Campbell-Selick formula, 81
Carlsson modules, 74, 79
Cartan-Serre basis, 69
Cartier isomorphism, 8
category, $\mathbb{K}$-linear, 21
coinvariants, 2
comodule structure, 30
comodules, 61
composition of two functors, 12
composition of two strict polynomial functors, 23
coordinate algebra, xviii, 28
counit of adjunction, 8
cross-effect, 117

de Rham complex, 7, 43
degree of a functor, 12, 93, 118
degree of a strict polynomial functor, 21
detection modulo nilpotents theorem, 52
diagonalizable, 16
divided powers algebra, 4
dual of a functor, 11

Euler formula, 8, 19
excess, 58, 61, 69
extension of scalars, 5
exterior algebra, 4

finite generation theorem, 47
formal group, 56
    additive, 56
Frobenius endomorphism, 2
Frobenius kernel, 29
Frobenius reciprocity, 31
Frobenius twist, 11, 23, 32
functor
    analytic, 91
    degree of a, 12, 21, 93, 118
    diagonalizable, 16
    finite degree, 117
    polynomial, 12, 93
    strict polynomial, 21, 40
functor cohomology, xvi
fundamental classes, 45

group algebra, xviii, 29

Hochschild complex, 37
homology
    MacLane homology, 116
    of a small category, 109, 114
Hopf algebra, 28, 56
hypercohomology spectral sequences, 18

induced module, 31
infinitesimal group scheme, 29
invariants, 2, 30

K-theory, stable, 109
Kan extension, 121
Koszul complex, 5, 43
Kuhn cancellation lemma, 95

L-H-S spectral sequence, 35

MacLane homology, 116
May spectral sequence, 37
Miller algebra, 77
Milnor basis, 58, 61

norm homomorphism, 3, 5

polynomial functor, 12
polynomial module, 39

Quillen stratification, 48

rational representation, 39

Schur algebra, xviii, 22, 39
sequence of functors, connected, 110
sequence of functors, universal, 110
socle, 97
Steenrod algebra, 58, 103
strict polynomial functor, 40
symmetric algebra, 3

tensor algebra, 3
tensor product of functors, 23
tensor product, pointwise, 11
type $(FP)_\infty$, 13

unit of adjunction, 8
unstable
    algebra, 70
    comodule, 61
    free unstable module, 65
    locally finite module, 72
    module, 63, 70
    nilpotent module, 73
    reduced module, 73

Verschiebung, 56

weight subspace, 34
Witt vectors, 38

Yoneda lemma, 11, 111

# INDEX TERMINOLOGIQUE

∂-suite spectrale, 110

Adams-Gunawardena-Miller, théorème d', 90
Adem, relations d', 67, 68
adjoints, 8
admissible, monôme, 69
admissible, suite, 69
algèbre
  à puissances divisées, 4
  de Boole, 90
  de groupe, xviii, 29
  de Hopf, 28, 56
  de Miller, 77
  de Schur, xviii, 22, 39
  de Steenrod, 58, 103
  extérieure, 4
  instable, 70
  symétrique, 3
  tensorielle, 3
algèbre des coordonnées, xviii, 28
analytique, foncteur, 91

base de Cartan-Serre, 69
base de Milnor, 58, 61
Boole, algèbre de, 90
Brown-Gitler, modules de, 74

Campbell et Selick, formule de, 81
Carlsson, modules de, 74, 79
Cartan-Serre, base de, 69
Cartier, isomorphisme de, 8
catégorie $\mathbb{K}$-linéaire, 21
classes fondamentales, 45
cohomologie des foncteurs, xvi
coinvariants, 2
comodule, structure de, 30
comodules, 61
composition de deux foncteurs, 12

composition de deux foncteurs polynomiaux
  stricts, 23
counité d'adjonction, 8

détection modulo les nilpotents, théorème de,
  52
de Rham, complexe de, 7, 43
degré d'un foncteur, 12, 93, 118
degré d'un foncteur polynomial strict, 21
diagonalisable, 16
dual d'un foncteur, 11

effet croisé, 117
Euler, formule d', 8, 19
excès, 58, 61, 69
extension des scalaires, 5

foncteur
  polynomial, 93
  analytique, 91
  de degré fini, 117
  degré d'un, 12, 21, 93, 118
  diagonalisable, 16
  polynomial, 12
  polynomial strict, 21, 40
Frobenius, endomorphisme de, 2
Frobenius, noyau de, 29
Frobenius, réciprocité de, 31
Frobenius, twist de, 11, 23, 32

génération finie, théorème de, 47
groupe affine, 60
groupe formel, 56
  additif, 56

Hochschild, complexe de, 37
homologie
  d'une petite catégorie, 109, 114
  de MacLane, 116

Hopf, algèbre de, 28
hypercohomologie, suite spectrale de, 18

induction, 31
instable
   algèbre, 70
   comodule, 61
   module, 63, 70
   module instable libre, 65
   module localement fini, 72
   module nilpotent, 73
   module réduit, 73
invariants, 2, 30

K-theory stable, 109
Kan, extension de, 121
Koszul, complexe de, 5, 43
Kuhn, lemme d'annulation de, 95

localement fini, module, 72
Lyndon-Hochschild-Serre, suite spectrale de, 35

MacLane, homologie de, 116
May, suite spectrale de, 37
Miller, algèbre de, 77
Milnor, base de, 58, 61
module instable, 63, 70
module instable libre, 65

nilpotent, module, 73
norme, 3, 5

poids, 34
polynomiale, représentation, 39
$p$-point abélien, 51
produit tensoriel de foncteurs, 11, 23

Quillen, stratification de, 48

réduit, module, 73
rationnelle, représentation, 39

schéma en groupes
   affine, 28
   infinitésimal, 29
Schur, algèbre de, xviii, 22, 39
socle, 97
sous-groupe à un paramètre, 49
Steenrod, algèbre de, 58, 103
suite connexe de foncteurs, 110
suite universelle de foncteurs, 110

type $(\mathrm{FP})_\infty$, 13

unité d'adjonction, 8

Verschiebung, 56

Witt, vecteurs de, 38

Yoneda, lemme de, 11, 111

# PANORAMAS ET SYNTHÈSES

**2003**

16. V. FRANJOU, E.M. FRIEDLANDER, T. PIRASHVILI, L. SCHWARTZ – *Polynomial functors, unstable modules and cohomology of finite group schemes*

15. M. BOILEAU, S. MAILLOT, J. PORTI – *Three-dimensional orbifolds and their geometric structures*

**2002**

14. P. DEHORNOY, I. DYNNIKOV, D. ROLFSEN, B. WIEST – *Why are braids orderable ?*

13. M. BABILLOT, R. FERES, A. ZEGHIB (avec la collaboration de E. BREUILLARD) – *Rigidité, groupe fondamental et dynamique (édité par P. FOULON)*

**2001**

12. N. GANTERT, J. GARNIER, S. OLLA, Z. SHI, A.-S. SZNITMAN – *Milieux aléatoires (édité par F. COMETS, É. PARDOUX)*

11. M. AUDIN, J.W. MORGAN, P. VOGEL, D. BENNEQUIN – *Nouveaux invariants en Géométrie et en Topologie (édité par F. DUMAS, J.-Y. LE DIMET, S. PAYCHA)*

**2000**

10. C. ANÉ, S. BLACHÈRE, D. CHAFAÏ, P. FOUGÈRES, I. GENTIL, F. MALRIEU, C. ROBERTO, G. SCHEFFER – *Sur les inégalités de Sobolev logarithmiques (avec une préface de D. BAKRY et M. LEDOUX)*

9. W. CRAIG – *Problèmes de petits diviseurs dans les équations aux dérivées partielles*

**1999**

8. D. CERVEAU, É. GHYS, N. SIBONY, J.-C. YOCCOZ (avec la collaboration de M. FLEXOR) – *Dynamique et géométrie complexes*

7. X. BUFF, J. FEHRENBACH, P. LOCHAK, L. SCHNEPS, P. VOGEL – *Espace de modules des courbes, groupes modulaires et théorie des champs*

**1998**

6. D. ALPAY – *Algorithme de Schur, espaces à noyau reproduisant et théorie des systèmes*

**1997**

5. C. KASSEL, M. ROSSO, V. TURAEV – *Quantum groups and knot invariants*

**1996**

4. M. ZINSMEISTER – *Formalisme thermodynamique et systèmes dynamiques holomorphes*

3. J. BERTIN, J.-P. DEMAILLY, L. ILLUSIE, C. PETERS – *Introduction à la théorie de Hodge*

2. C. VOISIN – *Symétrie miroir*

**1995**

1. S. TABACHNIKOV – *Billiards*

**1994**

0. J.-P. RAMIS – *Séries divergentes et théories asymptotiques*

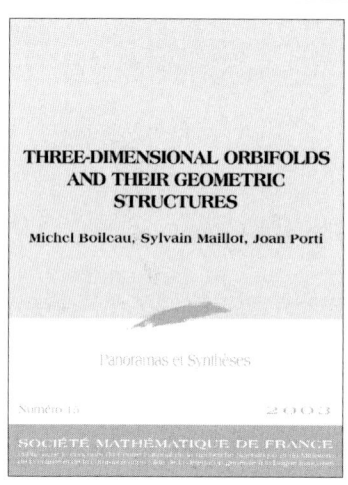

# Panoramas et Synthèses

# THREE-DIMENSIONAL ORBIFOLDS AND THEIR GEOMETRIC STRUCTURES

## Michel Boileau - Sylvain Maillot - Joan Porti

Orbifolds locally look like quotients of manifolds by finite group actions. They play an important rôle in the study of proper actions of discrete groups on manifolds. This monograph presents recent fundamental results on the geometry and topology of 3-dimensional orbifolds, with an emphasis on their geometric properties.

Une orbivariété est localement le quotient d'une variété par un groupe fini. Cette notion joue un rôle important dans l'étude des actions propres de groupes discrets sur les variétés. Cette monographie présente des résultats fondamentaux récents sur la géométrie et la topologie des orbivariétés de dimension 3, en mettant l'accent sur leurs propriétés géométriques.

**Prix public\* : 25 € ; Prix membre\* : 18 €**
**\* Frais de port non compris**

Commandes
Maison de la SMF, BP 67, 13274 Marseille Cedex 9 France
Tél : 04 91 26 74 64 - Fax : 04 91 41 17 51 - mail : smf@smf.univ-mrs.fr
url : http://smf.emath.fr/

# Einstein et la relativité générale
## Les chemins de l'espace-temps

## Jean Eisenstaedt

Comment, dans quel contexte, et au prix de quel effort la théorie de la relativité a vu le jour et évolué ? Cet ouvrage de vulgarisation nous donne le fil conducteur de cette aventure et associe intimement l'histoire des sciences et l'aspect biographique, en citant des journaux ou des correspondances d'astronomes ou de physiciens proches d'Einstein, découragés, enthousiastes ou même agressifs face à cette théorie difficile à accepter, à comprendre.

L'auteur insiste en particulier sur la « traversée du désert » d'Einstein, et sur la difficile institutionnalisation de la théorie. Les structures de la recherche en relativité sont restées longtemps artisanales ; il n'y a pas eu d'enseignement suivi sur la relativité avant les années 1950.

L'élaboration de la théorie, replacée dans le contexte de l'époque, est pour ainsi dire vécue de l'intérieur par le lecteur qui en découvre le développement heurté, sa croissance lente et son douloureux manque de résultats face à la théorie quantique.

On y comprend notamment comment les trous noirs, qui n'ont pu être posés ni pensés lors de la naissance de la théorie, vont être « inventés », compris, acceptés dans les années 1970... permettant une interprétation révolutionnaire de la théorie qui conduira au renouveau actuel.

**Collection CNRS Histoire des sciences**
**348 pages, 29 €**

Pour trouver et commander nos ouvrages :

**LA LIBRAIRIE de CNRS ÉDITIONS**, 151 bis, rue Saint-Jacques - 75005 PARIS
Tél. : 01 53 10 05 05 - Télécopie : 01 53 10 05 07 - Mél : librairie@cnrseditions.fr

**Site Internet : www.cnrseditions.fr**
Frais de port par ouvrage : France : 5 € - Etranger : 5,5 €

Pour plus de renseignements, n'hésitez pas à contacter
**le Service clientèle de CNRS ÉDITIONS**, 15, rue Malebranche - 75005 Paris
Tél : 01 53 10 27 07/09 - Télécopie : 01 53 10 27 27 - Mél : cnrseditions@cnrseditions.fr